Principles of Communication: A First Course in Communication

RIVER PUBLISHERS SERIES IN COMMUNICAITONS

Volume 6

Consulting Series Editor

Marina Ruggieri
University of Roma Tor Vergata
Italy

Other books in this series:

Volume 1
4G Mobile & Wireless Communications Technologies
Sofoklis Kyriazakos, Ioannis Soldatos, George Karetsos
September 2008
ISBN: 978-87-92329-02-8

Volume 2
Advances in Broadband Communication and Networks
Johnson I. Agbinya, Oya Sevimli, Sara All, Selvakennedy Selvadurai,
Adel Al-Jumaily, Yonghui Li, Sam Reisenfeld
October 2008
ISBN: 978-87-92329-00-4

Volume 3
Aerospace Technologies and Applications for Dual Use A New World of Defense and Commercial in 21st Century Security
General Pietro Finocchio, Ramjee Prasad, Marina Ruggieri
November 2008
ISBN: 978-87-92329-04-2

Volume 4
Ultra Wideband Demystified Technologies, Applications, and System Design Considerations
Sunil Jogi, Manoj Choudhary
January 2009
ISBN: 978-87-92329-14-1

Volume 5
Single- and Multi-Carrier MIMO Transmission for Broadband Wireless Systems
Ramjee Prasad, Muhammad Imadur Rahman, Suvra Sekhar Das, Nicola Marchetti
April 2009
ISBN: 978-87-92329-06-6

Principles of Communication: A First Course in Communication

Author

Kwang-Cheng Chen

*Graduate Institute of Communication Engineering
and Department of Electrical Engineering
National Taiwan University
Taipei 10617
Taiwan (ROC)*
chenkc@ieee.org

River Publishers

Routledge
Taylor & Francis Group
LONDON AND NEW YORK

Published 2009 by River Publishers
River Publishers
Alsbjergvej 10, 9260 Gistrup, Denmark
www.riverpublishers.com

Distributed exclusively by Routledge
4 Park Square, Milton Park, Abingdon, Oxon OX14 4RN
605 Third Avenue, New York, NY 10017, USA

First issued in paperback 2023

Principles of Communication: A First Course in Communication / by Kwang-Cheng Chen.

Routledge is an imprint of the Taylor & Francis Group, an informa business

Publisher's Note
The publisher has gone to great lengths to ensure the quality of this reprint but points out that some imperfections in the original copies may be apparent.

While every effort is made to provide dependable information, the publisher, authors, and editors cannot be held responsible for any errors or omissions.

ISBN 13: 978-87-7022-972-2 (pbk)
ISBN 13: 978-87-92329-10-3 (hbk)

Dedication

To our parents and families who believed in our education.

Preface

Global information communication technology (ICT) has driven tremendous applications and services in industry to benefit human being's life. The core of ICT lies in communications. The first course in communications has been listed as one of the courses that undergraduate students in electrical engineering must take and must select in global universities. There are a few well written textbooks as the first course of communications, including Communication Systems written by S. Haykins that has been used at the Department of Electrical Engineering, National Taiwan University. However, for this one-semester first course, those books reaching 700–800 pages are generally too much for many students within a semester. Under the motivation of a textbook fitting one semester, this book *Principles of Communications: A First Course* has been developed based on my nearly 10 years' teaching in the first course of communications at the National Taiwan University. My goal is to write a textbook with appropriate number of pages to address the fundamental theory for analog and digital communications, with introduction to modern digital communication technology.

During the tenure of the appointment as the Irving T. Ho chair professor, the author had more time to put together the lecture notes into this book. The first chapter describes some important milestones in communications and networks, which can not be a complete list. However, it might invoke students' motivation in devoting study and future research and development in communication technology. Chapters 2–6 establish the fundamental part as the first course of communications in university or colleague, with slight favor of mathematical treatments to ensure student's solid background in communication technology.

Chapters 7–10 represent some more advanced technology in modern digital communication systems such as CDMA used in 3G cellular and GPS, and OFDM in WiFi and WiMAX. In the mean time, in these four chapters, we try to avoid complicated mathematical development and try to introduce in more intuitive and more engineering approach so that those who just want to have common knowledge about technology behind modern communication systems can enjoy reading without struggling in mathematics. Of course, in order to understand modern communication technology in a solid way, typical textbooks at graduate level are required, such as *Digital Communications* authored by J. Proakis.

The rest of the book is organized as follows. Chapter 2 reviews random processes and filtering. Chapter 3 introduces analog communications that has been used for a century in broadcasting and communications. Chapter 4 briefs sampling of waveforms and pulse modulation. Chapter 5 is the core of fundamental communication theory to introduce optimal receiver concept. Chapter 6 develops passband communications that is commonly used in modern digital communications, wired or wireless. Related issues such as multiple access and link calculations are also introduced. For one-semester undergraduate course, Chapters 1–6 are considered mandatory. In case of two quarters and having more time to brief advanced materials in digital communications, a part of all from Chapters 7–10 can be used. Chapter 7 describes basic concepts of error correcting codes and error control, which is one of the major advantages to adopt digital communications. Chapter 8 briefs challenges in modern wireless communications, from modeling the channel to fading and diversity. Popular orthogonal frequency division multiplexing is oriented in Chapter 9. Spread spectrum communications and code division multiple access concepts are introduced in Chapter 10. These four chapters allow those who are not communication engineers to understand basics of modern communication techniques, without getting lost in complicated mathematics and special expertise regarding communication theory.

Last but definitely not the least, the author would like to thank many assistants in the past years who have helped in the preparation of course class note; support from the Irving T. Ho Memorial Foundation for their endowed chair at the National Taiwan University; mental support from his family (my Parents, wife Christine, daughter Chloe, and son Danny).

K.C. in Taipei, Taiwan

Contents

1 History and Milestones of Communication Technology   **1**

1.1 General Sense of Communications   1
1.2 History of Modern Communication   3
1.3 Communication Networks   7
1.4 Mobile Communication   11
1.5 Information and Communication Theory   14
1.6 Further Reading   16

2 Filtering of Random Processes and Signals   **19**

2.1 Probability and Random Variables   19
     2.1.1 Mathematical Development   19
     2.1.2 Useful Random Variables   20
2.2 Random Process   22
2.3 Filtering of Random Processes   27
2.4 Gaussian Random Processes   33
2.5 Noise and Its Representation   35

3 Analog Communications   **45**

3.1 Continuous Wave Modulation   45
3.2 Amplitude Modulation (AM)   46
3.3 Linear Modulation Schemes   49
3.4 Frequency Division Multiplexing (FDM)   57
3.5 Angle Modulation   58
3.6 Frequency Modulation   59
     3.6.1 Narrowband FM   61
     3.6.2 Wideband FM   63

3.7   Superheterodyne Receiver   72
3.8   Performance Analysis in Noise   73
      3.8.1   Coherent Detection in an AM Receiver   75
      3.8.2   Noise in AM with Envelope Detection   77
      3.8.3   Noise in FM Receivers   79
3.9   Phase Locked Loops   83

4 Pulse Modulation and Digital Coding   97

4.1   Sampling   99
4.2   Pulse-Amplitude Modulation (PAM)   102
4.3   Quantization   107
4.4   Time-Division Multiplexing (TDM)   113
4.5   Delta Modulation   116
4.6   Linear Prediction   118
4.7   Digital Coding of Waveforms   123

**5 Optimal Receiver of Digital Communication
    Systems   135**

5.1   Matched Filter   136
5.2   Linear Receiver   147
5.3   Optimal Linear Receiver   151
      5.3.1   Adaptive Equalization   156
      5.3.2   Line Coding   157
      5.3.3 M-ary PAM and Applications to
              Telecommunications   161
5.4   Signal Space Analysis   161
5.5   Correlation Receiver   172
5.6   Commonly Applied Signaling   184

6 Passband Digital Transmission   195

6.1   Digital Modulations   195
      6.1.1   Coherent PSK   197
      6.1.2   Coherent Frequency Shift Keying   205
      6.1.3   Non-coherent Digital Modulations   220
6.2   Spectral Efficient Modulations   227

6.3   Synchronization   230
    6.3.1   Open-Loop Spectral Line Generation
            Methodology   232
    6.3.2   Optimal Bit/Symbol Synchronization   232
    6.3.3   Optimum Carrier Phase Estimation   236
    6.3.4   Phase Locked Loop (PLL)   236
    6.3.5   Joint Estimation of Carrier and Symbol Timing   238
6.4   Link Calculations   238
6.5   Multiple Access   242
    6.5.1   Slotted Multi-access and ALOHA System   243
    6.5.2   Slotted ALOHA   244
    6.5.3   Unslotted ALOHA   245
6.6   Exercises   246

7   Error-Correcting Codes   **251**

7.1   ARQ   251
7.2   Linear Block Codes   252
7.3   Convolutional Codes   256
7.4   Trellis-Coded Modulation   260

8   Communications Over Wireless in Fading Channels   **267**

8.1   Channel Modeling   267
8.2   Bit-error-rate (BER) Analysis Over Fading Channels   271
8.3   Diversity   272

9   Orthogonal Frequency Division Multiplexing   **279**

9.1   Introduction   279
9.2   Synchronization for OFDM Signals   287
9.3   Channel Estiamtion and Equalization   290

**10 Spread Spectrum Communication and Code
   Division Multiple Access**   **293**

10.1 Baseband Equivalent SS Communications   294
    10.1.1 Baseband Equivalent Model   296
    10.1.2 Anti-Interference in Multi-Access   297
    10.1.3 Anti-multipath   297

10.2 DS-SS Communications   298

10.3 FH-SS Systems   305

10.4 CDMA and Multiuser Communications   307

    10.4.1 Spreading Codes   307

    10.4.2 Near–Far Problem   308

    10.4.3 Multiuser Detection (MUD)   309

Reference   **311**

Index   **319**

1

History and Milestones of Communication Technology

1.1 General Sense of Communications

One of the key features about intelligence is the ability of communication to exchange idea and thinking, which fundamentally distinguishes human beings from other species in the world. Communication also allows accumulation of experience and knowledge, as a fundamental to create civilization. To scientifically study the technology of enhancing the capability of communication, we can therefore define *communication* as transmitting some kind of *information* from one place (*transmitter* in engineering) to another (*receiver* in engineering). The medium to carry a signal between a transmitter and a receiver is known as *channel*. Speech communication among human beings (and other highly intelligent animals) is an example, if we consider language as a signal, mouth as a transmitter, and ear(s) as receiver(s). The challenge since thousands years ago is how to extend the scope of communication than speech communication in the range or in the number of audience (terminals in engineering). More examples of communication that we can think about include smokes made by Indians, waving flags for signals by scouts, navy, or ancient military commanders, horns/drums or bells to instruct some messages to a large group of people, etc.

Modern communication technology may trace back the root technology from the development of physics, especially the electro-magnetic (EM) theory in the 19th century, which might explain the origin of electrical engineering in scientific research. Up to today, modern communication technologies are all based on EM or optical foundation. We apply such EM methods to create signals representing certain meaning, to transmit such signals to designated receiver(s), and then to interpret such signals into messages. Figure 1.1 depicts an abstract

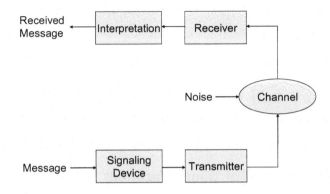

Fig. 1.1 Abstract model for communication.

model for communication, no matter it is a modern or an ancient way. For modern communication, we use a signaling device to transform a message into an electrical (or optical) signal representing this message. A transmitter sends the signal into the channel and a receiver collects the signal from the channel. There is added noise onto the signal in channel (in fact it can be at the transmitter and receiver circuits but we use a channel to summarize all such effects). Nonlinear distortion of the signal could occur in the channel, and we can model the channel as a filter to represent such a phenomenon. We mathematically model noise as a random process and usually model as an additive white Gaussian noise (AWGN), which will be characterized in Chapter 2 in detail. Once the receiver collects the right signal, the received message can be created via appropriate interpretation of an electrical/optical signal. In fact, such an abstract model can be applied to many other scenarios, such as read-write on the magnetic/optical disks, or even DNA copying and combination into another cell.

There are some common terminologies related to communication that we would like to introduce first. Ratio (without any unit) is very much needed to consider signal degradation in communication technology. We define r as R dB (decibel) where $R = 10 \log r$; we say 3 dB (more precisely, 3.01 dB) for two times and -3 dB for one-half. We can further measure power by the ratio with 1 mW and then take dB. For example, 0.1 W is 100 mW and is 20 dBm by taking dB for 100.

The ratio of signal power to noise power is known as signal-to-noise ratio and is commonly abbreviated as *SNR* or *S/N*.

1.2 History of Modern Communication

The commercial telegraph services were first started by W. Cooke and W. Wheatstone in England as early as 1834. In 1844, S. Morse who invented Morse codes (Figure 1.2 quoted from Wikipedia) that were widely used in telegraph until the late 20th century brought telegraph services in the United States. Atlantic cable was established between England and France in 1951. By 1861, an 18,000 km cable had been laid

INTERNATIONAL MORSE CODE

1. A dash is equal to three dots.
2. The space between parts of the same letter is equal to one dot.
3. The space between two letters is equal to three dots.
4. The space between two words is equal to five dots.

Fig. 1.2 Morse codes.

and a 5000 km cable actually worked all over the world. By mid-1870s, J. Stearns and T. Edison made it possible to transmit 2–4 telegraph signals over a single wire.

The major milestone in the history of communication technology is the invention of telephone by A. G. Bell in 1876. His patent was filed just a few hours earlier than E. Gray. Human voice communication could be carried over a much longer distance via a convenient way since then. In 1880, Bell Company had leased nearly 100,000 instruments. During the late 19th century, the major challenges of telephone were switching and long-distance transmission. Successful development had continued. A. B. Strowger patented an automatic dialing system to be used from 1892 to mid-1970s. G. Campell and M. Pupin filed a patent of inductive leading to make long-distance call possible in 1900. According to A. Fleming (England) and Lee de Forest (United States), AT&T built phone lines between New York and San Francisco. In addition, Earlang initiated the origin of queuing theory, a branch of mathematics and statistics, which impacts system engineering and telecommunication/computer networks in a significant manner. However, trans-Atlantic telephone cable was laid until 1950s.

Around the same period of time, radio communication was born and evolved. During 1860s, British scientist J. C. Maxwell developed the famous Maxwell equations in EM theory, aspiring a new branch of physics. When the application of EM theory became important, universities started to offer degrees in electrical engineering instead of treating as a special area in physics. The Darmstadt University of Technology founded the first chair and the first faculty of electrical engineering worldwide in 1882. In 1883, the Darmstadt University of Technology and Cornell University introduced the world's first course in electrical engineering, and in 1885 the University College London founded the first chair of electrical engineering in the United Kingdom. The University of Missouri subsequently established the first department of electrical engineering in the United States in 1886.

In 1888, German scientist H. Hertz generated EM radiation in his laboratory. The Famous Italian G. Marconi invented wireless signaling apparatus in 1896. Subsequently, radio signal from

Newfoundland was received in England in 1901, which created a new era of radio communication. Unfortunately, the useful values of radio communication were proven by three marine disasters in the early 20th century: White Star liner Republic in 1909, the well-known Titanic in 1912, and Volturno in 1913. Since 1920s, radio has been widely used for broadcasting.

Claude Shannon with Bell Laboratories published his pioneer paper "A Mathematical Theory of Communications" in 1947, which laid the foundation of digital communication theory and created a new area of information theory; more details are introduced in Chapter 7. Shockley, Bardeen, and Brattain invented transistors in 1948 in Bell Laboratories, which would realize digital technology for communications, computers, integrated circuits, and many others in the coming years. In the mean time, after revolution in modern mathematical analysis in the early 20th century, modern theory for probability and stochastic processes has been developed. Later innovations of communication theory and theoretical foundation of data networks are pretty much based on such advances. Chapter 4 will describe the basic statistical communication theory that was initially completed in 1960s to design the optimal receiver in digital communication.

Another major technology breakthrough for human being is the satellite into the space, which of course created a challenge for communication technology too. People quickly realized that satellites could be possibly used for cross-continent communications, though such an idea could be traced back hundreds of years. In 1950s, J. Pierce wrote "how satellite communication system might work." In 1957, Sputnik was sent into the earth's orbit by Russians, which sparked competition of space technology between the former USSR and the United States. The world's first communication satellite *"Echo I"* was successfully launched and operated in 1960. It is the so-called *passive* communication satellite to relay signals without amplification or any processing. In 1962, the first *active* communication satellite *"Telstar I"* was launched and its successors for generations are still serving today's trans-continent commercial communications. Figure 1.3 shows (a) Echo I, which is also known as "Early Bird", (b) state-of-the-art Telstar IX, and (c) earth station with a 100-feet dish antenna.

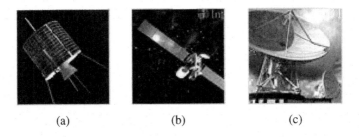

(a)   (b)   (c)

Fig. 1.3 Communication satellites: (a) Echo I; (b) Telstar IX; (c) Earth station.

Satellite communication and space communication are in fact using radio technology, which we generally call wireless communications. Note that wireless communications usually adopt radio but not always. A well-known exception is Infrared Data Association; although point-to-point file transfer between computers and consumer devices adopts infrared for transmission of signals, we can still consider infrared as an extremely high-frequency radio. Radio propagation suffers from many time-varying effects by Mother Nature and is difficult to predict and control, which constantly creates challenges for communication engineers. On the other hand, wires are less time-varying with rather stable transmission characteristics, at least in the order of bit or even frame duration. In addition to cooper wires and cables to guide EM waves carrying signals since the early stage of communication, to find a good technology to transmit signals through wires in long distance and in high bandwidth (i.e., data rates) is something favored. A new medium to allow effective optical transmission became a good idea, except typical glass to guide light waves suffered a great path loss to prohibit long-distance communication. However, such an idea was even more desirable after the invention of laser in 1959–1960, which made the operation at high-speed signaling (i.e., extremely a large volume of data transmission) possible.

In 1966, K. Charles Kao and G. A. Kockham proposed a clad glass fiber, of course with the purity of glass, as the waveguide for lightwave signals. Up to 1970, fiber loss reached 20 dB/km and I. Hayashi et al. demonstrated successful transmission using a semiconductor laser. Actual installation of optical fiber has happened in the mid-1970s.

Fig. 1.4 (a) 1000-core optical fiber, and (b) 8-core communication cable structure.

Optical fiber was used as a long-distance back-haul cable from early days, but it is the technology to end users to support high-bandwidth communication needs, such as FTTx (FTTH, fiber to the home; FTTB, fiber to the building). Via dense wavelength division multiplexing, 10 Gbps (bits per second) optical fiber is commercially available now and 100 Gbps is getting mature in the laboratory. Figure 1.4 depicts the photo of (a) a 1000-core optical fiber and (b) an 8-core communication cable structure.

The state-of-the-art communication technology for wires primarily focuses on the enhancement of existing communication cables/wires/lines capacity, such as xDSL (ADSL or HDSL to enhance the transmission capability of digital subscriber lines). The rationale behind this is the difficulty in the replacement of existing wired communication infrastructure, in terms of construction engineering and cost/time to pay. Chapter 10 will discuss the technologies for xDSL.

1.3 Communication Networks

Up to this point, we have discussed point-to-point communication. However, more desirable communication involves any two points in the group. As we mentioned that early telephone communication had a challenge in switching that builds up a connection between any two users inside the group of telephone users. Let us call this group as a telephone network and each user equipment (telephone) is a node inside the network. In modern communication, we not only have to deal with point-to-point communication but also have to deal with communication network among users.

For telephone networks, we may imagine some "physical connection circuits" to represent point-to-point communication links in such telecommunication networks. There exists a central switching system to exchange traffic in larger scale of circuit switching networks, no matter by operators in early days or by advanced digital electronic switching. It can be considered as "deterministic" multiplexing of signals inside a communication network. Circuit switching technology has migrated from hand-made, cross-bar to program controlled electronic switching system. The state-of-the-art telephone switching system is centrally hosted by a sophisticated computer and an extremely large program with billion lines of codes.

Along with the invention of computers, a new networking technology for computers as nodes came out, and we may call such networks as data networks or computer networks. The target information in computer networks is binary data in the form of "frames" or "packets", rather than voice in telephone networks. A node in computer networks generally has the capability of store-and-forward, and thus switching function is realized by *packet switching* instead of *circuit switching* in conventional telephone networks. Packet switching is also known as "stochastic" multiplexing of data packets due to the store-and-forward nature. Each data packet has a destination address in the header information, and a node reads such information to determine routing to the destination node. Central switching function likely does not exist anymore. Figure 1.5 is an example of modern computer communication network with store-and-forward network nodes, bridges, routers, gateways, local area networks (LAN), etc. Some major progress of data networks in early days includes L. Kleinrock's PhD dissertation "Information Flow in Large Communication Nets" at MIT in 1961 and then a book "Communication Nets" in 1964; P. Baran's "On Distributed Communications" in 1964; D. W. Watts considered packets and packet switching. An important step is ARPANET led by L. G. Roberts. ARPANET was originally developed by the Department of Defense in the United States, and later as a computer network for research

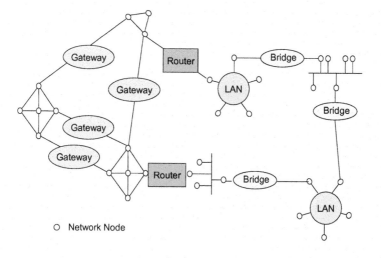

Fig. 1.5 Computer communication network.

institutions with users who were primarily conducting research in electrical engineering and computer science. R. Khan and V. Cerf paces the way from ARPANET to today's Internet. In 1978, V. Cerf et al. developed a transmission control protocol (TCP) and internet protocol (IP) from the early TCP work at Stanford University. Google hired V. Cerf as the "Chief Internet Evangelist" in September 2005.

It is difficult to define the starting point of Internet as migration from ARPANET. However, there still are some major technology milestones worth mentioning, along with successful commercial development. In the early 1990s, Netscape was developed as a tool for web browsing, and Microsoft's Internet Explore joined later. When the *World Wide Web* is a popular term among the fast-growing Internet users, Yahoo emerged as an Internet gateway for users. Routers and switches for the Internet core network and backbone give a new-dimension technology, which is different from that for the conventional telecommunication (phone) network. Internet also gives business a new era, and e-commerce has been a new business model in each segment of business and industry. We can consequently see i- and e- everywhere. Amazon for Internet book store and shopping, and eBay for Internet auction are among the well-known new business landscape. With more and more information on Internet, Stanford-graduate students created an Internet search engine technology and started a successful company Google. By Web 2.0, peer-to-peer networking services and information sharing entertains everyone on the Internet, such as YouTube for video-sharing services. Combining with mobile communication discussed later, we can now see m- and u- all over the world.

IEEE 802 is a project of the IEEE Computer Society (ironically, not Communications Society) to define the international standards for the International Standard Organization (ISO) in the areas of LANs and metropolitan area networks (MANs), which consists of the main part of computer networks to provide access from computers to the Internet. ISO defines an open system interconnection (OSI) architecture to allow inter-operability of different vendor equipment and devices. There are seven layers in OSI as shown in Figure 1.6, which will be carefully addressed in any textbook in computer networks or data networks. Two parts exist in data link control layer: logical link control (LLC) and medium access control (MAC, to/from the physical layer (PHY)).

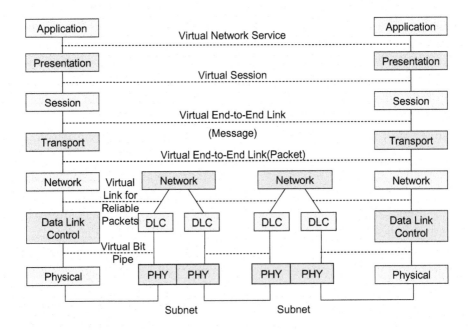

Fig. 1.6 OSI architecture.

PHY to reliably transmit/receive bits in the communication process is the major scope of this book.

The IEEE 802 deals with the lower two layers in LAN/MAN. With the commonly defined LLC, several working groups have been defining different sets of MAC and PHY. Some well-known examples include

- IEEE 802.3 Carrier Sense Multiple Access with Collision Detection (CD), which is the widely applied Ethernet.
- IEEE 802.5 Token-Ring.
- IEEE 802.11 Wireless LAN, which is also known as WiFi.
- IEEE 802.16 Wireless Broadband Access, which is also known as WiMAX.

Networking technology invokes a great deal of theoretical study such as queuing theory and traffic control in the networks. In fact, with a wide range of application of processors, computer networks exist everywhere. For example, each state-of-the-art BMW 7-series car has thousands of processors inside to work together supporting satisfactory functions from the sensor operation to the multimedia entertainment. Sensors using wireless communication described later can form networks with applications from security to body health monitoring. Routing information packets through appropriate paths to the destination is another challenge for intelligence, which surprisingly is a mathematical problem equivalent to the traveling salesman problem or bandit problem (what is your optimal strategy to win in front of slot-machines in casino). Massive parallel computing also involves communication and networking among processors. Operation research is therefore another type of close mathematical area with networking.

1.4 Mobile Communication

Another communication networking technology as new as the Internet might be mobile communication with radio communication and wireless networks together. One-way dispatch police system was in use in 1921. In 1946, AT&T provided mobile telephone services. However, we could see automatic trunked radio systems in 1960s. After FCC allocating 115 MHz spectrum (during 806–947 MHz) for land mobile communication, the first commercial cellular service was in the United States in 1983. Based on frequency division multiple access (FDMA), mandated AT&T AMPS was used in the United States. Along with other FDMA analog cellular systems, this is known as the first-generation wireless communication systems (or the first-generation cellular systems). Since the late 1980s, digital cellular technology had been developed as the second-generation (2G) wireless communication systems (cellular systems), based on time division multiple access. The most well-known example of such 2G systems might be global system for mobile communications (GSM), a pan-European standard. GSM is still the most widely used cellular technology today, with billions of users today. International Mobile Telecommunication-2000 (IMT-2000, or known as 3G, the third-generation wireless communication systems) had proposed technology deadline in June 1998. IMT-2000 now has two widely accepted versions (3GPP and 3GPP2) using frequency division multiplexing, and some versions based on time division multiplexing including time division-synchronize code division multiple access (TD-SCDMA) and a new possible version originated from mobile WiMAX Figure 1.7. 3GPP, 3GPP2, TD-SCDMA are all based on code vision multiple access (CDMA) technology.

In fact, the general scope of wireless communication contains two major types of communication systems in addition to traditional narrow-band communication (that is the major scope in Chapters 1–8 in this book): CDMA derived from spread spectrum communications and orthogonal frequency division multiplexing (OFDM). The world's first spread spectrum communication system was built up at MIT in 1920s. From 1950s to 1970s, MIT Lincoln Laboratories developed great efforts in technology about "communication by noise" with tremendous outcomes, and described a lot of them. Research at MIT and defense agencies had become open from 1970s. People had quickly got interests in multi-user communication to share the same spectrum at the same time by the spread spectrum communication technique, which is known as CDMA. M. Pursley first calculated multiple-access interference for such an *interference-limited* communication, while traditional narrow-band communication is either *power-limited* (i.e., communication link performance being dominated by transmission power such as satellite communication link, and increasing transmission power can enhance the link performance) or *bandwidth-limited* (i.e., communication link performance being dominated by signal bandwidth which is usually very valuable in many cases such as wireless/radio communication). To resolve such an interference-limited communication challenge, multi-user detection/communication was introduced by S. Verdu and V. Poor in 1980s, which can be applied to many multi-user and multi-terminal communication problems today and is till an active research area today. I. Jacobs and A. Viterbi funded Qualcomm in 1985 to realize CDMA cellular technology resulting IS-95 and to build up the foundation of 3G cellular (also known as IMT-2000), which uses the so-called wideband CDMA technology at 5 MHz and its multiple bandwidth.

In spite of sensitivity to channel variations, OFDM (a multi-carrier communication technology) has been primarily used in wireless local communications due to its advantages supporting high spectral efficiency. OFDM was originally proposed by R. Chang. After major improvements such as DFT realization and cyclic prefix, extension to multi-users as a new multiple access, OFDM-related technologies have been mostly adopted in the wireless standards of IEEE 802, such as

- IEEE 802.11a and IEEE 802.11g for wireless LANs, also ETSI (European Telecommunication Standard Institute) HIPERLAN (High-Performance Radio LANs).
- IEEE 802.11n adopting multi-antenna technology with OFDM as the most update version of wireless LANs.
- IEEE 802.15.3a and Bluetooth 3.0 using multi-band OFDM to meet FCC regulations of ultra-wide band communications.
- IEEE 802.16 using OFDM and orthogonal FDMA (OFDMA), while the OFDMA version of the IEEE 802.16e is also known as mobile WIMAX.

To enhance the receiver performance, multiple-antenna techniques have been employed since the last few years in the 20th century other than just conventional beamforming. Bell Laboratories researchers developed V-BLAST to leverage spatial multiplexing. Two dimensional Rake receiver was adopted into 3G cellular. Space–time codes to utilize both transmit diversity and receiver diversity was developed

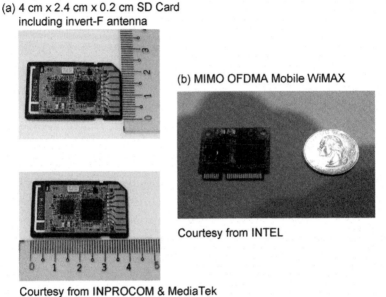

Fig. 1.7 (a) SD wireless LAN card with antenna, and (b) mobile WiMAX module with US Quarter as a reference size.

and widely used with OFDM(A) to achieve high spectral efficiency. Multi-input–multi-out (MIMO) coding and processing becomes a general terminology for such technologies and is widely designed into the state-of-the-art mobile communication systems.

1.5   Information and Communication Theory

The origin of information and communication theory might be C. Shannon's pioneer research to establish information theory described earlier in this chapter, which defines the channel capacity as follows:

$$C = W \log_2 \left(1 + \frac{S}{N} \right), \tag{1.1}$$

where C is the channel capacity under AWGN, W is the bandwidth for communication, and $\frac{S}{N}$ is the available SNR ratio at the receiver. We further define R as the (data) rate of communication, in the unit of bps. Information theory tells us that we can always find a way to achieve reliable communication (that is, error rate that measures digital communication system performance can approach zero), if $R \leq C$. Otherwise, if $R > C$, reliable communication is not possible and thus error rate for such communication systems is always bounded away from zero. This has been proven by the random coding technique; however, the method(s) to achieve channel capacity in designing a communication system has not been described. It takes quite a few decades for human knowledge to approach this theoretical limit. The development of telephone line modem can brief such efforts. Data communication has a major application to use telephone line (twist pair) for modem to connecting personal computers and other digital data facilities. The lines are for voice communication around 0–4 kHz bandwidth. We usually do not want to use low-frequency portion of this 0–4 kHz and consequently we expect around 2.4 kHz spectrums available for data transmission. With an operating SNR at 30 dB, by (1.1), we can estimate 24 kHz bps as the channel capacity. Consequently, a 19.2 kbps modem is around 1 dB away from theoretical limit. Of course, later by compression technology, we can use 54 kbps modem prior to broadband technologies widely available.

Early advance in data communications adopted equalization and then adaptive equalization pioneered by R. Lucky et al. at Bell Laboratories. By using modulation such as binary phase shift keying (PSK) and quadrature PSK, 300, 600, 1.2 kbps, ... modems were developed. It appears to be a bandwidth-limited communication case. We have to increase the spectral efficiency of modulations and error-correcting codes. We usually apply block codes and convolutional codes as error-correcting codes to provide extra protection of digital modulation/signaling against channel noise and impairments, at the expense of extra bandwidth. For example, we encode 4 information bits into 7 bits through some algebraic manipulations, then into the channel. The receiver can decode these 7 bits into 4 information bits with more tolerance of errors. Convolutional codes have been known to have good performance against errors. Its optimal decoding methodology was developed by A. Viterbi, which was later known as Viterbi algorithm and widely used in various aspects in communication theory. It is also applied into general sense of communication such as storage problems, while modern CD technology adopts concatenated codes consisting of Reed-Solomon codes and convolutional codes with interleaving to protect stored information from scratch. In fact, error-correcting codes have been widely used not only in communication but also in computer industry.

In the late 1970s, an IBM Zurich Lab. researcher G. Ungenboeck noted that considering both modulation and coding at the same time, instead of treating them separately as common ways, can improve the overall system performance. In 1982, he finally completed the famous *set-partitioning principle* to effectively combine modulation and coding. For example, using quadrature amplitude modulation and industrial standard convolutional codes, 3 dB extra coding gain can be easily achieved by proper mapping of coded bits and signal constellations. Such a technology, (trellis) coded modulation, was quickly applied to voice-band modem to reach 9.6 kbps and higher data rate later. Lee-Fang Wei later extended trellis coded modulation (TCM) into rotation-invariant and further into multi-dimensional TCM so that 19.2 kbps voice-band modem could be realized. By numerical example for Equation (1.1), it is only 1 dB away from theoretical bound, as the first time for human being to approach the limit in communications. Such a gap to theoretical limit has been further narrowed in 1990s by developing *turbo codes* through iterative decoding of two convolutional codes, or by revisiting *low-density parity check codes* originally invented by MIT professor Robert Gallager.

Advance of communication technology and theory always interact with the advance of circuit technology such as semiconductor. In light of Morse' Law, the new frontier of communication technology may move along with nano-science, such as quantum communications, molecular or biological communications, for even higher data rate or even lower power consumption.

1.6   Further Reading

Almost all the above major technology breakthroughs described can be found as technical papers in the IEEE journals, while the IEEE Xplore provides digital library access. A few books specialized in one or few subjects of the above description are listed as follows, as a complementary reading beyond the scope of this introductory communication book.

- Digital communications [11] [63] [83] [116]
- Information theory [33] [46]
- Error-correcting/control codes [111]
- Synchronization and receiver design [48] [52] [75] [101]
- Detection and Estimation [113] [156]
- CDMA and Multi-user communications [153] [157] [160]
- OFDM [29]
- Mobile communications [3] [28] [56] [122]
- Optical communications [77]
- Data Networks [16]

Of course, the above-mentioned list is not exhaustive. Many great books and articles are subject to your exploration.

2

Filtering of Random Processes and Signals

2.1 Probability and Random Variables

Probability has been intuitively used since thousands years ago, which we usually consider as classical probability. After advances in modern analysis in the early 20th century, modern probability theory and thus stochastic process theory have been developed, which is pretty much complete for modern engineering to apply in 1960s, although still a state-of-the-art mathematics is in progress to have an impact on many branches of engineering, especially in communications and networks.

Although strict mathematical theory to deal with probability requires tools such as measure theory,[1] we intend to introduce concepts based on Riemann Integral at the level of undergraduate calculus. In the mean time, our purpose is to supply readers fundamental knowledge in this book. More in-depth understanding may require appropriate mathematical tools and development beyond this level.

2.1.1 Mathematical Development

Modern probability theory is established using the concept of probability space, which considers three elements in developing the theory: sample space, event, and probability measure.

Definition 2.1. A *probability space* (Ω, B, P) is a triple to consisting of a sample space Ω, a event space B of subsets of Ω, and a probability measure P to assign a real number to every member A_i of B.

---

[1] Interested engineering readers may read a typical textbook such as *Probability, Random Processes, and Ergodic Properties* authored by R. M. Gray and L. D. Davisson, or any book at similar level.

Then we construct the probability measure based on the following axioms:

Axiom 2.1. $1 \geq P(A_i) \geq 0$.

Axiom 2.2. $P(\Omega) = 1$.

Axiom 2.3. *If A_i are mutually disjoint, then $P(\cup_i A_i) = \sum_i P(A_i)$.*

Axiom 2.1 implies that the probability of any event falls within $[0, 1]$, that is, non-negative and not greater than 1. Axiom 2.2 implies that the total probability is 1, which can provide a natural bound for Axiom 2.1 (thus, Axiom 2.1 can be without an upper bound). Axiom 2.3 starts from the fact that the probability of union of two disjoint events is the sum of probabilities of these two events.

After defining probability, we are moving to the development of random variables.

Definition 2.2. A *random variable* defined on (Ω, B) and taking values in (A, B_A) is a mapping or function $f : \Omega \to A$ with the property that if $A_i \in B_A$, then $f^{-1}(A_i) = \{\omega : f(\omega) \in A_i\} \in B$.

Note that a random variable is in fact a deterministic mapping (function) from events to real numbers. Its name, random variable, is in fact very misleading. Firstly, random variable is not random, but deterministic. Second, random variable is not a variable, but is a function or more precisely a mapping.

2.1.2 Useful Random Variables

The readers of this book are expected to learn elementary probability based on Calculus. Therefore, we summarize a few useful random variables in communications and networks and explain their relationship from intuitions in this subsection.

In modern probability theory, we have two categories of random variables and later on random processes: discrete and continuous, which rely on a simple discrepancy: countable or non-countable. Countable can be finite or countably infinite.

Example 2.1 (Countably Infinite and Uncountable). We consider real numbers in $[0, 1]$. The rational numbers (for those that can be represented by the form p/q, where p and q are integers) are infinite but countable. However, the irrational numbers are infinite but uncountable.

Some well-known discrete random variables with their probability density functions (pdf), $P(X = m) = f(m)$, include

- Geometric:

$$f(m) = (1 - p)p^m, \quad 1 > p > 0 \quad \text{and} \quad m = 0, 1, 2, \dots. \quad (2.1)$$

- Binomial:

$$f(m) = \binom{M}{m}(1 - p)^{M-m}p^m,$$

$$1 > p > 0, \quad M \in N, \quad \text{and} \quad m = 0, 1, 2, \dots, M. \quad (2.2)$$

- Poisson: with parameter λ

$$f(m) = \frac{\lambda^m}{m!}e^{-\lambda}, \quad m = 0, 1, 2, \dots. \quad (2.3)$$

There are more well-known continuous random variables, and we summarize their pdf as follows:

- Uniform:

$$f(x) = \frac{1}{b - a}, \quad x \in [a, b]. \quad (2.4)$$

- Exponential: with parameter λ

$$f(x) = \lambda e^{-\lambda x}, \quad x \geq 0. \quad (2.5)$$

- Gaussian: with mean μ and variance σ^2

$$f(x) = \frac{1}{\sqrt{2\pi\sigma^2}}e^{-(x-\mu)^2/2\sigma^2}, \quad -\infty \leq x \leq \infty. \quad (2.6)$$

- Normal: if a Gaussian random variable has zero mean and variance 1, it has a normal distribution.

- Gamma: with parameter λ and degree of n

$$f(x) = \lambda \frac{(\lambda x)^{n-1}}{(n-1)!} e^{-\lambda x}, \quad x > 0. \tag{2.7}$$

- Chi-square: in (2.7), we may define $\Gamma(n) = (n-1)!$; therefore

$$f(x) = \frac{x^{n/2-1}}{2^{n/2}\Gamma\left(\frac{n}{2}\right)} e^{-x/2}, \quad x \geq 0. \tag{2.8}$$

Note some interesting relationships among some random variables, which would be useful in the future to study communications and networks. Exercise 3 shows that the Gamma distribution as in (2.7) is the sum of n independent and identically distributed exponential random variables. Exercise 4 shows that the chi-square distribution (2.8) is that of $Y = X^2$, where X has a normal distribution.

2.2 Random Process

Once we establish random variables, we are ready to define random processes, which would be fundamentally used in communications and networks. Intuitively, a *random process* is a collection of random variables, usually with time index. In other words, a random process has two dimensions of ensemble and time axis, or We may call it as "indexed ensemble of random variables." If such an index is countable, then it is a discrete random process. If the index is uncountable, then it is a continuous random process.

Definition 2.3. A *discrete-time random process* is a sequence of random variables $\{X_i\}_{i \in I}$ or $\{X_i; i \in I\}$, where I is an index set, defined on a common probability space (Ω, B, P).

Definition 2.4. A *continuous-time random process* is a collection of random variables $\{X_t\}_{t \in T}$ or $\{X_t; t \in T\}$ defined on a common probability space (Ω, B, P), where T is uncountable.

We may consider Figure 2.1 as an illustration of random process. Mapping from a sample space, we may have different outcome traces

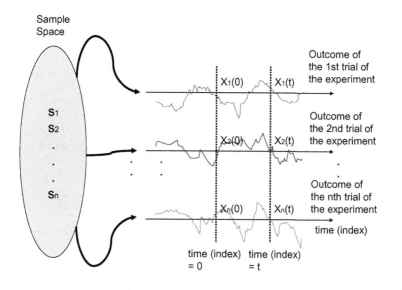

Fig. 2.1 Ensemble traces of a sample function for a random process.

for a random process, with one axis for time and another for ensemble. Different from a random variable whose outcome of the random experiment is mapped into a number, the outcome of the random experiment for a random process is mapped into a function with (time) index.

In the engineering world, we often observe a stable statistical characterization of a process independent of time during the observation period. Such a process is said to be *stationary*. Otherwise, it is called *non-stationary*.

Definition 2.5. A random process $X(t)$ is *stationary in strict sense* (or *strictly stationary*) if the joint distribution function is time-invariant, that is,

$$F_{X(t_1+\tau),\ldots,X(t_k+\tau)}(x_1,\ldots,x_k) = F_{X(t_1),\ldots,X(t_k)}(x_1,\ldots,x_k).$$

Similarly, two random processes $X(t)$ and $Y(t)$ are *jointly strictly stationary* if the joint finite-dimensional distributions of two sets of random variables $X(t_1),\ldots,X(t_k)$ and $Y(t'_1),\ldots,Y(t'_j)$ are invariant with respect to the time origin $t = 0$, $\forall k, j$, and all combinations of observation times $t_1,\ldots,t_k$ and $t'_1,\ldots,t'_j$.

Definition 2.6. For a strictly stationary random process $X(t)$, its mean is given by

$$\mu_X(t) = E[X(t)] = \int_{-\infty}^{\infty} x f_{X(t)}(x)\, dx. \tag{2.9}$$

Since the pdf of $X(t)$ is independent of time, the mean of a strictly stationary process is a constant, that is,

$$\mu_X(t) = \mu_X, \quad \forall t. \tag{2.10}$$

Definition 2.7. The *autocorrelation function* of the process $X(t)$ is

$$R_X(t_1, t_2) = E[X(t_1)X(t_2)]$$
$$= \int_{-\infty}^{\infty} \int_{-\infty}^{\infty} x_1 x_2 f_{X(t_1)X(t_2)}(x_1, x_2)\, dx_1\, dx_2. \tag{2.11}$$

We may consider an interesting case between times 0 and $(t_2 - t_1)$ for $t_2 > t_1$. Since $X(t)$ is strictly stationary, (2.11) may imply that the autocorrelation function of a strictly stationary process depends only on the time difference $t_2 - t_1$, that is,

$$R_X(t_1, t_2) = R_X(t_2 - t_1), \quad \forall t_1, t_2. \tag{2.12}$$

The *autocovariance function* of a strictly stationary process $X(t)$ is therefore defined as

$$C_X(t_1, t_2) = E[(X(t_1) - \mu_X)(X(t_2) - \mu_X)]$$
$$= R_X(t_2 - t_1) - \mu_X^2. \tag{2.13}$$

Definition 2.8. A random process $X(t)$ satisfying (2.10) and (2.12) is *wide-sense stationary*.

This definition is widely used in communications, signal processing, and system engineering. We may simply refer $X(t)$ as a *stationary process.* For easy notation, the autocorrelation function of a stationary process $X(t)$ can be re-defined as

$$R_X(\tau) = E[X(t + \tau)X(t)], \quad \forall t. \tag{2.14}$$

The autocorrelation function has the following important properties:

Property 1.

   (a) The mean-square value of the process $E[X^2(t)] = R_X(0)$.
   (b) $R_X(\tau)$ is an even function of τ, that is, $R_X(\tau) = R_X(-\tau)$.
   (c) $R_X(\tau)$ has its maximal magnitude as $\tau = 0$, that is, $|R_X(\tau)| \le R_X(0)$.

Example 2.2. To prove Property 2.9(c), we consider the non-negative quantity $E[(X(t + \tau) - X(t))^2] \ge 0$. Then, we can obtain $2R_X(0) - 2R_X(\tau) \ge 0$.

Example 2.3 (Sinusoidal Wave with Random Phase). It is common to consider a sinusoidal waveform with the random phase as

$$X(t) = A \cos(2\pi f_c t + \Theta), \tag{2.15}$$

where A and f_c are constants and Θ is a random variable uniformly distributed over $[-\pi, \pi]$. That is,

$$f_\Theta(\theta) = \frac{1}{2\pi}, \quad -\pi \le \theta \le \tau. \tag{2.16}$$

$X(t)$ is usually used to represent the locally generated carrier at the receiver of a communication system and to demodulate the received waveform. The random phase Θ denotes the phase difference between locally generated carrier and the carrier in the received waveform transmitted from the transmitter. The autocorrelation function of $X(t)$ is given by

$$R_X(\tau) = E[X(t + \tau)X(t)]$$
$$= E[A^2 \cos(2\pi f_c t + 2\pi f_c \tau + \Theta) \cos(2\pi f_c t + \Theta)]$$

$$= \frac{A^2}{2} E[\cos(4\pi f_c t + 2\pi f_c \tau + 2\theta)] + \frac{A^2}{2} E[\cos(2\pi f_c \tau)]$$

$$= \frac{A^2}{2} \int_{-\pi}^{\pi} \frac{1}{2\pi} \cos(4\pi f_c t + 2\pi f_c \tau + 2\theta) d\theta + \frac{A^2}{2} \cos(2\pi f_c \tau)$$

$$= \frac{A^2}{2} \cos(2\pi f_c \tau). \tag{2.17}$$

Now, we consider a more general case of two random processes $X(t)$ and $Y(t)$ with autocorrelation functions $R_X(t,u)$ and $R_Y(t,u)$. The *cross-correlation functions* of $X(t)$ and $Y(t)$ are given by

$$R_{XY}(t,u) = E[X(t)Y(u)]$$
$$R_{YX}(t,u) = E[Y(t)X(u)]. \tag{2.18}$$

We can in fact use a matrix form to represent the correlation properties of two random processes as

$$\mathbf{R}(t,u) = \begin{bmatrix} R_X(t,u) & R_{XY}(t,u) \\ R_{YX}(t,u) & R_Y(t,u) \end{bmatrix}. \tag{2.19}$$

In case $X(t)$ and $Y(t)$ are jointly stationary, then (2.19) can be reduced to

$$\mathbf{R}(\tau) = \begin{bmatrix} R_X(\tau) & R_{XY}(\tau) \\ R_{YX}(\tau) & R_Y(\tau) \end{bmatrix}. \tag{2.20}$$

Property 2. For a pair of stationary processes $X(t)$ and $Y(t)$,

(a) $R_{XY}(\tau) = R_{YX}(-\tau).$ \hfill (2.21)

(b) $|R_{XY}(\tau)| \leq \dfrac{R_X(0) + R_Y(0)}{2}.$ \hfill (2.22)

Before the end of this section, recall the intuitive definition of a random process as a collection of random variables, which introduces two dimensions to observe a random process: ensemble and time. Communication systems are designed based on the observation of random waveforms after channel perturbations. Let us consider a sample function $x(t)$ of a stationary process $X(t)$ during the observation interval $-T \leq t \leq T$.

Definition 2.9. The random process is *ergodic*, if

$$\lim_{T \to \infty} \mu_x(T) = \lim_{T \to \infty} \frac{1}{2T} \int_{-T}^{T} x(t)\, dt = \mu_X = \int_{-\infty}^{\infty} x f_X(x)\, dx. \quad (2.23)$$

Here, we adopt the definition by considering only expectation, that is, *ensemble average equal to time average.* In fact, "ergodic" can also be defined in autocorrelation function. Almost all random processes that we encounter in communications and networks (or engineering systems) are ergodic, which provides us a great deal of flexibility in derivations and explorations of system nature. However, we have to mention that the desirable mathematic definition of ergodic should be $P(A) = 1$ or 0, for every time invariant set A.

2.3   Filtering of Random Processes

The fundamental part of system science is to explore signal nature; we usually observe from the time domain. In communications, it is also important to "observe" from the frequency domain. Frequency domain is rather abstract and is constructed from mathematics. An intuitive way to consider frequency domain behaviors of signals and systems is to treat frequency domain as variations in time domain. Signal behaviors in time domain and frequency domain are linked through Fourier transform and Fourier series, so that we can precede frequency domain analysis of signals and systems.

Definition 2.10 (Fourier Series). Let the signal $x(t)$ be a periodic signal with period T. If (Dirichlet condition)

(a) $\int_{0}^{T} |x(t)|\, dt < \infty,$

(b) $\max\limits_{0 \le t \le T} |x(t)| < \infty,$

(c) The number of discontinuities of $x(t)$ in each period is finite, then $x(t)$ can be expanded as

$$x(t) = \sum_{n=-\infty}^{\infty} x_n e^{j2\pi \frac{n}{T} t}. \qquad (2.24)$$

$$x_n = \frac{1}{T} \int_0^T x(t) e^{-j2\pi \frac{n}{T} t} \, dt. \qquad (2.25)$$

Note (2.24) in fact forms an orthonormal expansion of a function in functional analysis. $\{e^{j2\pi \frac{n}{T} t}\}_{n=-\infty}^{\infty}$ can be re-written as $\{e^{j2\pi \omega t}\}_{n=-\infty}^{\infty}$ in some cases and serves as the orthonormal basis (i.e., the basis to satisfy both orthogonal property and normalization).

For general continuous waveforms, we have to use Fourier transform to serve the purpose.

Definition 2.11 (Fourier Transform). If $x(t)$ satisfies Dirichlet conditions, then its Fourier transform is defined as

$$X(f) = \int_{-\infty}^{\infty} x(t) e^{-j2\pi ft} \, dt. \qquad (2.26)$$

The original waveform can be obtained from inverse Fourier transform as

$$x(t) = \int_{-\infty}^{\infty} X(f) e^{j2\pi ft} \, df. \qquad (2.27)$$

We summarize a few important properties of Fourier transform and its applications as follows, while more details of Fourier transform can be found in any textbook related to signals and systems.

Property 3. If we denote the Fourier transform by F and the inverse Fourier transform by F^{-1}, then

(a) Linear property: $\mathsf{F}[\alpha x(t) + \beta y(t)] = \alpha X(f) + \beta Y(f).$

(b) Duality property: If $X(f) = \mathsf{F}[x(t)]$, then

$$x(f) = \mathsf{F}[X(-t)] \quad \text{or} \quad x(-f) = \mathsf{F}[X(t)].$$

(c) Time shift property: $\mathsf{F}[x(t - \tau)] = e^{-j2\pi f \tau} \mathsf{F}[x(t)]$.

(d) Scaling property: $\mathsf{F}[x(at)] = \frac{1}{|a|} X\left(\frac{f}{a}\right)$.

(e) Convolution property: $\mathsf{F}[x(t) * y(t)] = X(f)Y(f)$.

(f) Modulation property:

$$\mathsf{F}[x(t)\cos(2\pi f_0 t)] = \frac{1}{2}[X(f - f_0) + X(f + f_0)].$$

(g) Parseval's property: $\int_{-\infty}^{\infty} x(t)y^*(t)\,dt = \int_{-\infty}^{\infty} X(f)Y^*(f)\,df$.

(h) Rayleigh's property: $\int_{-\infty}^{\infty} |x(t)|^2\,dt = \int_{-\infty}^{\infty} |X(f)|^2\,df$, which suggests that the total energy in time domain is equal to the total energy in frequency domain.

(i) Autocorrelation property: $\mathsf{F}[R_X(t)] = |X(f)|^2$.

(j) Differentiation property: $\mathsf{F}\left[\frac{d}{dt}x(t)\right] = (j2\pi f)X(f)$.

(k) Integration property: $\mathsf{F}[\int_{-\infty}^{t} x(\tau)\,d\tau] = \frac{X(f)}{j2\pi f} + \frac{1}{2}X(0)\delta(f)$.

(l) Moment property: $\int_{-\infty}^{\infty} t^n x(t)\,dt = \left(\frac{j}{2\pi}\right)^n \frac{d^n}{df^n} X(f)|_{f=0}$.

The above-mentioned properties can be proved without major difficulty (see Exercise 6). Now, we are considering a random process $X(t)$ as the input of a linear time-invariant filter of impulse response $h(t)$ whose frequency response is $H(f)$. The output $Y(t)$ is also a random process, which is given by

$$Y(t) = X(t) * h(t) = \int_{-\infty}^{\infty} h(\tau)X(t - \tau)\,d\tau. \qquad (2.28)$$

The mean of $Y(t)$ is expressed as

$$\mu_Y(t) = E[Y(t)] = E\left[\int_{-\infty}^{\infty} h(\tau)X(t - \tau)\,d\tau\right]$$

$$= \int_{-\infty}^{\infty} h(\tau)E[X(t - \tau)]\,d\tau = \mu_X(t - \tau)\int_{-\infty}^{\infty} h(\tau)\,d\tau. \qquad (2.29)$$

Note that we may consider expectation as a linear operator $E \equiv \int f(x)\,dx \equiv \int dF(x)$ in derivations. When $X(t)$ is stationary,

$$\mu_Y = \mu_X \int_{-\infty}^{\infty} h(\tau)\,d\tau = \mu_X H(0), \tag{2.30}$$

where $H(0)$ is the DC (zero frequency) response of the system. Note that (2.30) pretty much coincides with our intuition. We may derive the autocorrelation function of $Y(t)$ in a similar manner by treating expectation as an operator such as integral

$$R_Y(t,u) = E[Y(t)Y(u)]$$

$$= E\left[\int_{-\infty}^{\infty} h(\xi)X(t-\xi)\,d\xi \int_{-\infty}^{\infty} h(\sigma)X(u-\sigma)\,d\sigma \right]$$

$$= \int_{-\infty}^{\infty} h(\xi)\,d\xi \int_{-\infty}^{\infty} h(\sigma)\,d\sigma\, E[X(t-\xi)X(u-\sigma)]$$

$$= \int_{-\infty}^{\infty} h(\xi)\,d\xi \int_{-\infty}^{\infty} h(\sigma)\,d\sigma\, R_X(t-\xi, u-\sigma). \tag{2.31}$$

When the input process $X(t)$ is stationary, the autocorrelation function of $X(t)$ is only a function of time difference between observation times $t-\xi$ and $u-\sigma$. By setting $\tau = t - u$, we have

$$R_Y(\tau) = \int_{-\infty}^{\infty} \int_{-\infty}^{\infty} h(\xi)h(\sigma)R_X(\tau - \xi + \sigma)\,d\xi\,d\sigma. \tag{2.32}$$

We may conclude from this derivation that if the input to a linear time-invariant filter is a stationary process, then the output is also a stationary process. Using (2.32), we can derive the mean value of the output process as follows:

$$E[Y^2(t)] = R_Y(0) = \int_{-\infty}^{\infty} \int_{-\infty}^{\infty} h(\xi)h(\sigma)R_X(\sigma - \xi)\,d\xi\,d\sigma. \tag{2.33}$$

Now, let us examine the frequency domain behaviors of the system by recalling $h(t)$ and $H(f)$ to be a Fourier transform pair. We would like to consider the second-order property of signals (or spectral analysis of a random process) in what follows.

Definition 2.12. The *power spectral density* (or *power spectrum*) of the stationary process $X(t)$ is given by

$$S_X(f) = \int_{-\infty}^{\infty} R_X(\tau)e^{-j2\pi f\tau}\, d\tau. \tag{2.34}$$

Also, $R_X(\tau)$ and $S_X(f)$ form a Fourier transform pair.

Power spectral density (psd) is a very fundamental nature in dealing with signals and systems. We may intuitively consider psd as the power distribution in the frequency domain (or power density at certain frequency). The integral of psd of the entire frequency range (such as $(-\infty, \infty)$) is the total power with respect to a signal.

Theorem 2.1. While $|H(f)|$ is the magnitude response of the linear time-invariant filter,

$$E[Y^2(t)] = \int_{-\infty}^{\infty} |H(f)|^2 S_X(f)\, df. \tag{2.35}$$

It suggests that the mean-square value of the output of a linear time-invariant filter with a stationary input process is equal to the entire integral frequency range of the psd of the input process multiplied by the squared magnitude of filter response in frequency domain.

Property 4.

(a) The DC term of the power spectral density for a stationary random process is equal to the total area under the curve of the autocorrelation function, that is,

$$S_X(0) = \int_{-\infty}^{\infty} R_X(\tau)\, d\tau. \tag{2.36}$$

(b) The mean-square value of a stationary process is equal to the total area under the curve of psd, that is,

$$E[X^2(t)] = \int_{-\infty}^{\infty} S_X(f)\, df. \tag{2.37}$$

(c) The psd of a stationary process is non-negative, that is,

$$S_X(f) \geq 0, \quad \forall f. \tag{2.38}$$

(d) The psd of a real-valued random process is an even function, that is,

$$S_X(f) = S_X(-f). \tag{2.39}$$

(e) The normalized psd has properties like a pdf of

$$p_X(f) = \frac{S_X(f)}{\int_{-\infty}^{\infty} S_X(f)\,df}. \tag{2.40}$$

Example 2.4. Continuing Example 2.3, the autocorrelation function has been derived as in (2.17). By taking Fourier transform, we can have

$$S_X(f) = \frac{A^2}{4}[\delta(f - f_c) + \delta(f + f_c)]. \tag{2.41}$$

Note the unitary feature of δ function to suggest the total area in (2.41) as $A^2/2$, which means total power.

Example 2.5 (Modulating a Random Process over a Sinusoidal Process). Continuing from Example 2.3, we form a new random process as follows:

$$Y(t) = X(t)\cos(2\pi f_c t + \Theta). \tag{2.42}$$

The random phase in (2.42) is uniformly distributed over $[-\pi, \pi]$ or equivalently over $[0, 2\pi]$. $X(t)$ and Θ usually come from independent sources, and thus we usually treat them statistically independent,

$$
\begin{aligned}
R_Y(\tau) &= E[Y(t+\tau)Y(t)] \\
&= E[X(t+\tau)\cos(2\pi f_c t + 2\pi f_c \tau + \Theta)X(t)\cos(2\pi f_c t + \Theta)] \\
&= E[X(t+\tau)X(t)]E[\cos(2\pi f_c t + 2\pi f_c \tau + \Theta)\cos(2\pi f_c t + \Theta)] \\
&= R_X(\tau)\frac{1}{2}E[\cos(2\pi f_c \tau) + \cos(4\pi f_c t + 2\pi f_c \tau + 2\Theta)] \\
&= \frac{\cos(2\pi f_c \tau)}{2}R_X(\tau). \tag{2.43}
\end{aligned}
$$

Its power spectral density is therefore given by

$$S_Y(f) = \frac{1}{4}[S_X(f - f_c) + S_X(f + f_c)]. \qquad (2.44)$$

Note that the above equation means that the psd of the random process $X(t)$ is being shifted to frequency components $\pm f_c$. We will explain this important behavior later in communications as *modulation*.

Theorem 2.2. Let $S_Y(f)$ be the psd of the output process $Y(t)$ by passing $X(t)$ into a linear time-invariant filter with impulse response $H(f)$ in frequency domain, that is,

$$S_Y(f) = |H(f)|^2 S_X(f). \qquad (2.45)$$

Equation (2.45) has an engineering meaning that the psd of the output process $Y(t)$ is equal to the psd of the input process $X(t)$ multiplied by the squared magnitude of frequency response of the filter. We may easily reach further observations. First, the DC term of a random process can be found using (2.36). Second, we may conclude that

$$E[X^2(t)] = R_X(0) = \int_{-\infty}^{\infty} S_X(f)df. \qquad (2.46)$$

It suggests that the energy does not change whenever it is evaluated in frequency domain or in time domain. These discussions also hold for time-average autocorrelation function known as *periodogram*, which is discussed in-depth in many graduate-level textbooks for Digital Signal Processing.

2.4 Gaussian Random Processes

The purpose of introducing random processes into the study of communication systems is to model the random behaviors from communication channels. The most widely applied model is through Gaussian random processes. Let us first recall Gaussian random variable with mean μ and variance σ^2, whose pdf is given by

$$\frac{1}{\sqrt{2\pi\sigma^2}} \exp\left(-\frac{(x-\mu)^2}{2\sigma^2}\right).$$

Gaussian random variable has an insight to look at. It is the most "random" random variable. If we evaluate the *entropy* (defined in Information Theory) of random variables, Gaussian random variable has the largest "entropy." The special case of Gaussian is a normal distribution with $\mu = 0$ and $\sigma^2 = 1$, which is useful in many cases of normalization in statistics (say, normalization of examination scores from a well-attended test).

We usually denote the Gaussian distribution with mean μ and variance σ^2 as $\mathsf{G}(\mu, \sigma^2)$ and the normal distribution as $\mathsf{N}(0,1)$.

Theorem 2.3 (Central Limit Theorem). $X_n, n = 1, 2, 3, \ldots, N$, denote a set of random variables that are statistically independent and identically distributed (that is, iid) with mean μ_X and variance σ_X^2. Define

$$S_N = \frac{1}{\sqrt{N}} \sum_{n=1}^{N} \frac{X_n - \mu_X}{\sigma_X}.$$

Then,

$$\lim_{N \to \infty} S_N = \mathsf{N}(0,1). \tag{2.47}$$

It pretty much explains the importance of Gaussian random variable; we can mix a large number of iid random variables to approach the normal distribution. Before discussing Gaussian processes, we introduce the joint Gaussian random variables that might not be independent.

Property 5 (Multivariate Gaussian Distribution). Let $\mathbf{X} = (X_1, X_2, \ldots, X_N)$ be a random vector of Gaussian random variables with mean vector $\mu = (\mu_1, \mu_2, \ldots, \mu_N)$ and covariance matrix $\mathbf{Q}$. The joint distribution of $\mathbf{X}$ is given by

$$f_{\mathbf{X}}(\mathbf{x}) = \frac{1}{(2\pi)^{N/2}|\mathbf{Q}|^{1/2}} e^{-\frac{1}{2}(\mathbf{x}-\mu)^T \mathbf{Q}^{-1}(\mathbf{x}-\mu)}, \tag{2.48}$$

where $\mathbf{x} = (x_1, x_2, \ldots, x_N)$.

Now, it is straightforward to define Gaussian random process as a collection of Gaussian random variables. Gaussian process is the most common candidate to model unknown noise in communication channels, and we will use in the following chapters and maybe in many future more advanced communication courses. Let us investigate more about Gaussian process and from linear filtering of a Gaussian process.

Property 6

   (a) If a Gaussian process is stationary, then it is strictly stationary.

   (b) If Gaussian random variables are sampled from a Gaussian process at indexed times, then they are jointly Gaussian with distribution (2.48).

   (c) If Gaussian random variables are sampled from a Gaussian process at indexed times and they are uncorrelated, then they are also statistically independent.

   (d) If a Gaussian process is applied as the input of a stable linear filter, then the output process of the filter is also Gaussian.

These properties are very useful when we characterize noise process in communications later.

There is another random process known as Poisson process (also known as point process or counting process) that is very useful in modeling traffic of communication networks. Poisson process can relate Poisson random variable (distribution of number of events), exponential random variable (inter-arrival time between two adjacent events), and Gamma random variables (time between multiple events) together. We will discuss about these later.

2.5 Noise and Its Representation

In communication systems, noise is used to designate all unwanted sources to disturb the transmission, processing, and receiving of signals. For example, *shot noise* arises due to the discrete nature of current flows, typically in photo-detector for optical communications that we will orient in the later chapter. For electronic communications, random behaviors from circuits and channel medium can be described by *thermal noise* whose power is the multiplication of noise temperature and *Boltzmann* constant. Note that noise temperature is not room temperature and is an equivalent value.

We usually start from noise analysis of communication systems based on a rather ideal model called *white noise* (see Figure 2.2). The psd of white noise modeled as a random process $W(t)$ is given by

$$S_W(f) = \frac{N_0}{2}. \tag{2.49}$$

By thermal noise model,

$$N_0 = kT_e, \tag{2.50}$$

where k is the *Boltzmann* constant, and T_e is the equivalent noise temperature. The autocorrelation function of white noise is given by

$$R_W(\tau) = \frac{N_0}{2}\delta(\tau). \tag{2.51}$$

White noise is named after its psd to cover the entire spectrum. However, it cannot be realistic due to its infinite total power. A more realistic example might be white noise after a low-pass filtering with bandwidth B, while Figure 2.3 illustrates such a case.

In fact, the receiver in a communication system always does some pre-processing of received waveform, which results in an overall

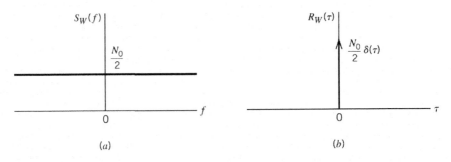

Fig. 2.2 Characterization of white noise: (a) power spectral density and (b) autocorrelation [66].

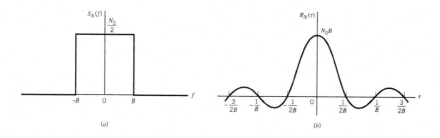

Fig. 2.3 Low-pass filtered white noise.

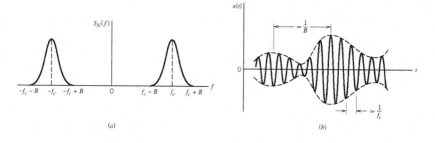

Fig. 2.4 Narrowband noise after receiving filter.

narrowband filtering (receiving filter) near the carrier frequency to
cause a narrowband noise prior to detection of signals (Figure 2.4).
White noise can thus be a good model.

Such a narrowband noise can be specified in two common
approaches:

- Orthogonal basis: in-phase (I) noise and quadrature-phase
  (Q) noise.
- Spherical basis: envelope and phase.

Definition 2.13 (Canonical Form of Narrowband Noise). Consider a narrowband noise $n(t)$ with two-side bandwidth $2B$ centered at
frequency f_c. We may represent $n(t)$ as

$$n(t) = n_I(t)\cos(2\pi f_c t) - n_Q(t)\sin(2\pi f_c t). \qquad (2.52)$$

Note that we may include $\sqrt{2}$ into the orthogonal basis in (2.52) for the purpose of normalization, that is,

$$n(t) = n_I(t)\sqrt{2}\cos 2\pi ft - n_Q(t)\sqrt{2}\sin 2\pi ft. \qquad (2.53)$$

Based on earlier section about Gaussian process, we can reach the following property:

Property 7. The in-phase component $n_I(t)$ and quadrature-phase component $n_Q(t)$ of a narrowband noise

    (a) have zero mean;
    (b) are jointly Gaussian;
    (c) are jointly stationary;
    (d) have the same power spectral density;
    (e) have the same variance as the narrowband noise $n(t)$;
    (f) the cross-correlation density of $n_I(t)$ and $n_Q(t)$ is purely imaginary;
    (g) if the narrowband noise $n(t)$ is Gaussian and its psd is symmetric about mid-band frequency f_c, then $n_I(t)$ and $n_Q(t)$ are independent.

In communication theory and system design, we consider the channel noise to be additive to a signal waveform and to be a Gaussian process and white noise, which is commonly known as additive white Gaussian noise (AWGN). The projection of AWGN onto orthogonal basis like canonical form forms iid Gaussians (as the original narrowband noise $n(t)$) from the property.

We may also represent noise in terms of envelope and phase components as follows:

$$n(t) = r(t)\cos[2\pi f_c t + \psi(t)], \qquad (2.54)$$

where $r(t)$ is the envelope and $\psi(t)$ is the phase and are given by

$$r(t) = \left[n_I^2(t) + n_Q^2(t)\right]^{1/2} \qquad (2.55)$$

and

$$\psi(t) = \tan^{-1}\left[\frac{n_Q(t)}{n_I(t)}\right]. \tag{2.56}$$

As $n_I(t)$ and $n_Q(t)$ are independent Gaussian random variables of zero mean and variance σ^2, their joint probability is given by

$$f_{N_I,N_Q}(n_I,n_Q) = \frac{1}{2\pi\sigma^2}\exp\left(-\frac{n_I^2 + n_Q^2}{2\sigma^2}\right). \tag{2.57}$$

Thus,

$$f_{N_I,N_Q}(n_I,n_Q)\,dn_I\,dn_Q = \frac{1}{2\pi\sigma^2}\exp\left(-\frac{n_I^2 + n_Q^2}{2\sigma^2}\right)dn_I\,dn_Q. \tag{2.58}$$

Define the transformation

$$n_I = r\cos\psi \tag{2.59}$$

$$n_Q = r\sin\psi \tag{2.60}$$

and thus

$$dn_I\,dn_Q = r\,dr\,d\psi. \tag{2.61}$$

Let R and Ψ be the random variables from the observation of $r(t)$ and $\psi(t)$:

$$\frac{r}{2\pi\sigma^2}\exp\left(-\frac{r^2}{2\sigma^2}\right)dr\,d\psi.$$

The joint pdf is given by

$$f_{R,\Psi}(r,\psi) = \frac{r}{2\pi\sigma^2}\exp\left(-\frac{r^2}{2\sigma^2}\right). \tag{2.62}$$

The phase random variable shall be uniformly distributed as

$$f_\Psi(\psi) = \begin{cases} \dfrac{1}{2\pi}, & 0 \le \psi \le 2\pi \\ 0, & \text{elsewhere.} \end{cases} \tag{2.63}$$

Then,

$$f_R(r) = \begin{cases} \dfrac{r}{\sigma^2} \exp\left(-\dfrac{r^2}{2\sigma^2}\right), & r \geq 0 \\ 0, & \text{elsewhere.} \end{cases} \tag{2.64}$$

Note that the above equation is Rayleigh distributed. We may use the following change of variables:

$$v = \frac{r}{\sigma}$$

$$f_V(v) = \sigma\, f_R(r).$$

The normalized form of (2.64) can be expressed as

$$f_V(v) = \begin{cases} v \exp\left(-\dfrac{v^2}{2}\right), & v \geq 0 \\ 0, & \text{elsewhere.} \end{cases} \tag{2.65}$$

Now, we can add a sinusoidal wave into the narrowband noise, whose sample function can be expressed as

$$x(t) = A\cos(2\pi f_c t) + n(t). \tag{2.66}$$

Using the canonical form, we can re-write it as

$$x(t) = n_I'(t)\cos(2\pi f_c t) - n_Q(t)\sin(2\pi f_c t), \tag{2.67}$$

where

$$n_I'(t) = A + n_I(t). \tag{2.68}$$

We assume $n(t)$ to be Gaussian with zero mean and variance σ^2. By Property 2.20, the joint density function is expressed as

$$f_{N_I', N_Q}(n_I', n_Q) = \frac{1}{2\pi\sigma^2} \exp\left(-\frac{(n_I' - A)^2 + n_Q^2}{2\sigma^2}\right). \tag{2.69}$$

Using envelope and phase representation, we have

$$r(t) = \left\{ [n_I'(t)]^2 + n_Q^2(t) \right\}^{1/2}$$

and

$$\psi(t) = \tan^{-1}\left[\frac{n_Q(t)}{n_I'(t)}\right].$$

Similar to Rayleigh derivations, we have

$$f_{R,\Psi}(r,\psi) = \frac{r}{2\pi\sigma^2} \exp\left(-\frac{r^2 + A^2 - 2Ar\cos\psi}{2\sigma^2}\right). \qquad (2.70)$$

To obtain the distribution of R, we average over possible values of phase

$$f_R(r) = \int_0^{2\pi} f_{R,\Psi}(r,\psi)\,d\psi$$

$$= \frac{r}{2\pi\sigma^2} \exp\left(-\frac{r^2 + A^2}{2\sigma^2}\right) \int_0^{2\pi} \exp\left(\frac{Ar}{\sigma^2}\cos\psi\right)d\psi. \qquad (2.71)$$

The integral on the right-hand side involves the *modified Bessel function of the first kind of zero order*, which is given by

$$I_0(x) = \frac{1}{2\pi} \int_0^{2\pi} \exp(x\cos\psi)\,d\psi. \qquad (2.72)$$

Substituting $x = Ar/\sigma^2$, we have

$$f_R(r) = \frac{r}{\sigma^2} \exp\left(-\frac{r^2 + A^2}{2\sigma^2}\right) I_0\left(\frac{Ar}{\sigma^2}\right). \qquad (2.73)$$

Define new variables,

$$v = \frac{r}{\sigma}$$

$$a = \frac{A}{\sigma}$$

and

$$f_V(v) = \sigma f_R(r).$$

The normalized form of *Rician* distribution is given by

$$f_V(v) = v \exp\left(-\frac{v^2 + a^2}{2}\right) I_0(av). \qquad (2.74)$$

Here we introduce both Rayleigh and Rician random variables, which are very useful in wireless communications and other communication systems. Generally and intuitively speaking, Rayleigh implies weak signal with a good amount of noise terms, but Rician implies a strong signal term embedded in noise terms.

Exercises

1. Derive the pdf of the Gamma distribution with parameters n and λ.

2. Exponential random variable is also known as the *memoryless* distribution, which can be used to describe the half-time in atomic radiation. Explain and prove this feature.

3. Show that the Gamma distribution as in (2.7) is the sum of n independent and identically distributed exponential random variables.

4. Show that the chi-square distribution (2.8) is that of $Y = X^2$, where X has a normal distribution.

5. Prove Property 2.10.

6. Prove Property 2.14.

7. Consider a random binary waveform consisting of a sequence of random bits "1"s and "0"s represented by values of $+A$ and $-A$, respectively. Show that

   (a) The autocorrelation function of this random process has a triangular waveform

   $$R_X(\tau) = \begin{cases} A^2\left(1 - \frac{|\tau|}{T}\right), & |\tau| < T \\ 0, & |\tau| \geq T. \end{cases}$$

   (b) The psd of this process is

   $$S_X(f) = A^2 T \operatorname{sinc}^2(fT).$$

8. Prove Theorem 2.1.

9. X is a Gaussian random variable with mean μ and variance σ^2. The random process $Y(t)$ is defined as

   $$Y(t) = X\cos(2\pi f_c t).$$

   Find the marginal density of $Y(t)$.

10. Let $X(t)$ be a zero-mean, stationary, Gaussian process with autocorrelation function $R_X(\tau)$.

    (a) What is the power spectral density $S_X(f)$?

(b) By observing $X(t)$ at some time t_k, what is the probability density function in this observation? This process is applied to a square-law device, which is defined by the input–output relation $Y(t) = X^2(t)$, where $Y(t)$ is the output.

(c) Show that the mean of $Y(t)$ is $R_X(0)$.

(d) Show that the autocorrelation function of $Y(t)$ is $2R_X^2(\tau)$.

3

Analog Communications

3.1 Continuous Wave Modulation

In communication, the signal is usually at baseband frequency range. We therefore use a continuous waveform generated by the transmitter as a *carrier* to carry information-bearing signals to the receiver of destination. The carrier is generally at a much higher frequency (denoted by f_c) than signals. Such a process to embed information and to translate signal into higher frequency range centered at f_c is called *modulation*. There are many reasons to use modulation from the beginning of electronic communications.

- Electromagnetic wave propagation.
- Multiplexing.
- Feasibility of antenna length.

Figure 3.1 shows a general block diagram of continuous wave modulation and demodulation.

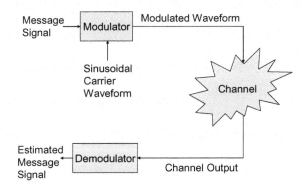

Fig. 3.1 Components of continuous wave modulation: (a) transmitter and (b) receiver.

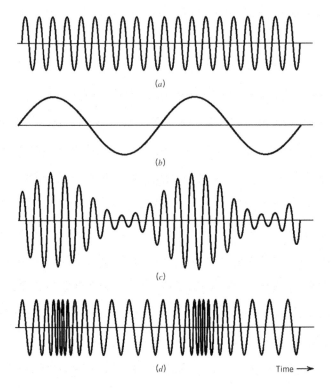

Fig. 3.2 (a) Carrier waveform; (b) signal waveform; (c) signal embedded into the amplitude; (d) signal embedded into the frequency.

As we explained earlier, there are three places where we can embed information messages into the waveform: amplitude, frequency, and phase (or angle). Figure 3.2 illustrates some intuitive waveform embedded signals in amplitude and in frequency.

3.2   Amplitude Modulation (AM)

For continuous wave modulation, we consider a sinusoidal carrier wave $c(t)$ as follows:

$$c(t) = A_c \cos(2\pi f_c t), \tag{3.1}$$

where A_c is the carrier amplitude and f_c is the carrier frequency. AM is the process to embed information to the amplitude of carrier waveform $c(t)$, which varies linearly with the baseband message $m(t)$

around the mean value $c(t)$.

$$s(t) = A_c[1 + k_a m(t)] \cos 2\pi f_c t. \tag{3.2}$$

where k_a is a constant called amplitude sensitivity of the modulator to generate modulated signal $s(t)$.

From Figure 3.3(b), the modulated waveform basically has the same envelope as baseband message $m(t)$, provided

(a) $\qquad |k_a m(t)| < 1 \quad$ for all t $\hfill (3.3)$

Figure 3.3(c) shows an example to violate this condition, which is known as over-modulated to exhibit phase reversals and thus envelope distortion.

(b) $\qquad f_c \gg W,$ $\hfill (3.4)$

where W is called message bandwidth. If (3.4) does not hold, the envelope cannot be visualized satisfactorily to cause difficulty in successful detection.

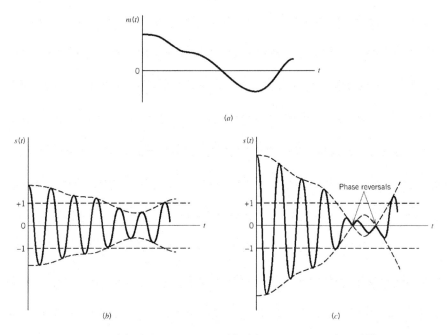

Fig. 3.3 AM process: (a) baseband signal $m(t)$; (b) AM wave for $|k_a m(t)| < 1$ for all t; (c) AM wave for $|k_a m(t)| > 1$ for some t.

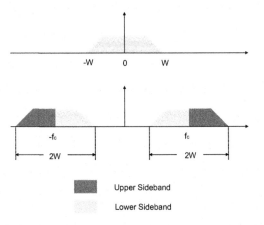

Fig. 3.4 Spectrum of an AM waveform.

In case the message $m(t)$ has Fourier transform $M(f)$ that is band limited to $[-W, W]$, we can have the Fourier transform of AM waveform $s(t)$ as

$$S(f) = \frac{A_c}{2}[\delta(f - f_c) + \delta(f + f_c)] + \frac{A_c k_a}{2}[M(f - f_c) + M(f + f_c)].$$
(3.5)

The spectrum consists of two versions of the baseband message spectrum translated to $\pm f_c$ with a scaled factor of $k_a A_c / 2$ in amplitude and two spikes centered at $\pm f_c$. Figure 3.4 depicts the *upper sideband* and *lower sideband* of a signal waveform. If we just look at the positive frequency portion of Figure 3.4, the transmission bandwidth for AM is $2W$.

From the spectrum of AM waveforms, we can clearly observe some drawbacks:

   (a) A portion of the power is carried in $\frac{A_c}{2}[\delta(f + f_c) + \delta(f - f_c)]$, which does not carry any information at all. It suggests that some power is wasted, where AM is power limited (communication link performance depends on power and power is the performance "bottleneck" of the system).
   (b) We may note symmetrical upper sideband and lower sideband, which suggests bandwidth wasted.

Finally, it is interesting to note that AM is still applicable nowadays. The major reason is not AM itself. AM is used for broadcasting at high-frequency (HF) band, which can leverage the reflection by the ionizing layer above the atmosphere for extremely long-distance propagation, such that even BBC from UK Europe can be heard in Asia for the other side of the earth beyond any possible light of sight.

3.3 Linear Modulation Schemes

Linear modulation can be defined in a canonical form as

$$s(t) = s_I(t)\cos(2\pi f_c t) - s_Q(t)\sin(2\pi f_c t). \tag{3.6}$$

We may extend AM into three types of linear modulations:

- *Double sideband-suppressed carrier* (DSB-SC) modulation: both upper sideband and lower sideband of $s(t)$ are transmitted, and the carrier is suppressed in the mean time.
- *Single sideband* (SSB) modulation: only one sideband (either upper or lower) is transmitted.
- *Vestigial sideband* (VSB) modulation: only a vestige of one sideband and a correspondingly modified version of another sideband are transmitted.

The DSB-SC is usually realized by a product modulator as shown in Figure 3.5(a). We can express the waveform as

$$s(t) = A_c m(t)\cos(2\pi f_c t). \tag{3.7}$$

The advantage of DSB-SC is obvious without carrier in the original AM. However, as we can observe from Figure 3.5(c), possible phase reversal is possible. The Fourier transform of $s(t)$ is given by

$$S(f) = \frac{A_c}{2}[M(f - f_c) + M(f + f_c)]. \tag{3.8}$$

To recover the baseband signal $m(t)$ from the DSB-SC waveform, a coherent detector as shown in Figure 3.6 can be employed, with an important feature of a locally generated sinusoidal waveform multiplied with the received waveform, then passing a low-pass filter (LPF).

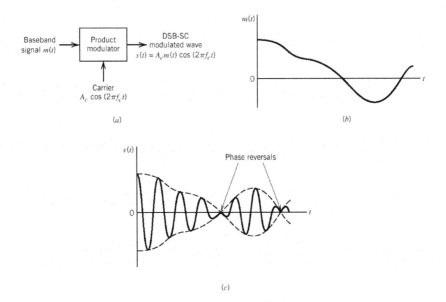

(a)

(b)

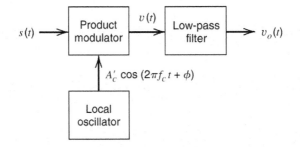

(c)

Fig. 3.5 (a) Product modulator; (b) baseband signal; (c) DSB-SC waveform.

Fig. 3.6 A coherent detector for DSB-SC.

The local oscillator delivers the signal waveform as $A'_c \cos(2\pi f_c t + \phi)$, where ϕ denotes the arbitrary phase difference measured with respect to the transmitted carrier. The product modulator output is therefore expressed as

$$v(t) = A'_c \cos(2\pi f_c t + \phi)s(t)$$
$$= A_c A'_c \cos(2\pi f_c t) \cos(2\pi f_c t + \phi)m(t)$$
$$= \frac{A_c A'_c}{2} \cos(4\pi f_c t + \phi)m(t) + \frac{A_c A'_c}{2}(\cos\phi)m(t). \qquad (3.9)$$

After the LPF it is given by

$$v_0(t) = \frac{A_c A_c'}{2}(\cos\phi)m(t). \tag{3.10}$$

The demodulated signal is therefore proportional to the message $m(t)$ when the phase error ϕ is a constant. Clearly, the demodulated signal reaches its maximum when $\phi = 0$ and becomes zero when $\phi = \pm\pi/2$, which is known as *quadrature null effect* of the coherent detector.

There are some commonly used terminologies in communication systems and are useful in the future:

- When we are talking about *coherent* communications or *coherent* receiver/demodulator, here we mean the existence of a local reference signal that is locally generated at the receiver. There is no need to recover phase, which is the same definition as in optical communications;
- When we are talking about *synchronous* communication/receiver, or even *coherent* communication/receiver in radio communications, we usually mean that the receiver must *exactly* recover the phase, in addition to frequency and timing. Otherwise, we call *noncoherent* (or asynchronous) communication or non-coherent demodulator/detector if we do NOT need to exactly recover the phase.

Also note that filter (either LPF or band-pass filter (BPF)) is usually required after any sort of signal processing, as undesired signal components will also be generated due to nonlinear effects from signal processing, and may further introduce undesirable situation such as *inter-modulation*.

Example 3.1 (Inter-Modulation). Suppose that we have a signal with three frequency components $f - (0.98)\varphi, f, f + (0.99)\varphi$ with $0 < \varphi \ll f$. We pass this signal into a cubic processing (that is, 3rd-order filtering). Within the frequency range of $[f - \varphi, f + \varphi]$, it is supposed to have only three frequency components. However, we in fact have frequency components at $f - (0.99)\varphi, f - (0.98)\varphi, f - (0.01)\varphi, f, f + (0.01)\varphi, f + (0.97)\varphi, f + (0.99)\varphi$ due to filtering effect.

Even ignoring the effects on amplitude, we can clearly observe extra frequency components; such a situation is known as *inter-modulation* created by signal components inside the signal bandwidth.

A famous receiver structure shown in Figure 3.7 is *Costas receiver*, which is widely used in modern analog and digital communication systems. The received waveform is split into two channels: *in-phase* channel (or I-channel) and *quadrature-phase* channel (Q-channel). Note that the phase discriminator (or phase detector (PD)), voltage-controlled oscillator (VCO), and mixer, to form a *phase locked loop* (PLL), and to be introduced later in this book. PLL is a fundamental part of communication systems and is applied to many other schemes even outside the scope of communications, such as processor, interfaces, etc. Figure 3.8 depicts a digital PLL (DPLL) with equivalent digital realization.

On the basis of the I-Q channel realization, we can build up a new system architecture of *quadrature-carrier multiplexing* or *quadrature*

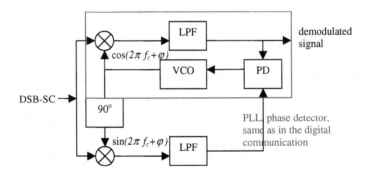

Fig. 3.7 Costas receiver.

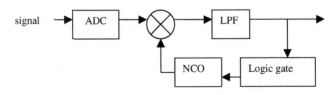

Fig. 3.8 DPLL.

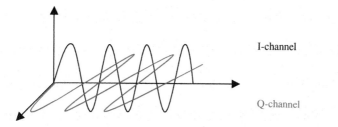

Fig. 3.9 Quadrature-carrier multiplexing.

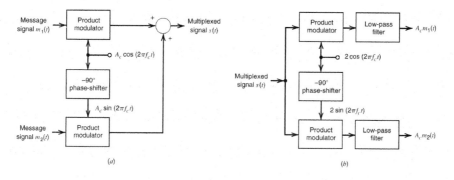

Fig. 3.10 QAM system: (a) transmitter; (b) receiver.

AM (QAM). Figure 3.9 depicts the intuitive signal representation to show the orthogonality between I-channel signal and Q-channel signal. Separate signals can be independently transmitted over these I-channel and Q-channel. The advantage is obvious when we double the communication bandwidth. However, after channel propagation, signals in original I, Q channels are likely no longer orthogonal, as a new challenge to design the communication system. The block diagram of quadrature-carrier multiplexing is illustrated in Figure 3.10.

For SSB modulation, either the upper sideband or the lower sideband is transmitted. As shown in Figure 3.11, we can use band-pass filtering to pass one of the sidebands from the waveform generated by DSB-SC product modulator. It also suggests that we can multiplex two signals for upper and lower sidebands, respectively.

We use a coherent detector as the typical SSB demodulator. We have to note that this demodulation assumes perfect synchronization between the transmitter and the receiver. Here, synchronization means

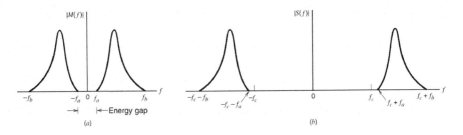

Fig. 3.11 Spectrum of an SSB modulation.

alignment of carrier and locally generated signal, which requires that

- A (low-power) pilot carrier is transmitted accompanying the selected sideband. This pilot can help synchronization at the receiver.
- A highly stable oscillator at the receiver is tuned to the same frequency as the carrier frequency.

Example 3.2 (Precision of the Oscillator). We usually measure the precision of an oscillator in terms of ppm, which means relative error at the order of 10^{-6}. A 100 ppm oscillator at 100 MHz means that the maximal frequency error $100 \times 10^6 \times 100 \times 10^{-6} = 10 \, \mathrm{KHz}$.

For VSB modulation, one of the sidebands is partially suppressed and a vestige of another sideband is transmitted to compensate that suppression. Frequency discrimination is usually used to generate a VSB-modulated waveform as follows:

- To design a VSB filter. Figure 3.12 illustrates an example with a vestige of the lower sideband.
- To use a product modulator generating DSB-SC waveform and then to pass through a BPF designed in the earlier step as shown in Figure 3.13.

In order to design the VSB filter as shown in Figure 3.12, the cut-off portion of the frequency response around the carrier frequency f_c exhibits *odd symmetry*. In other words, the following two conditions

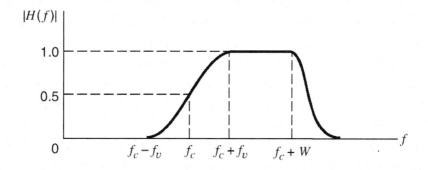

Fig. 3.12 Positive frequency magnitude response of VSB filter.

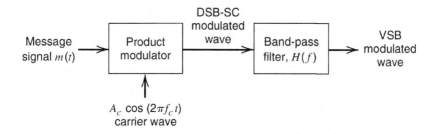

Fig. 3.13 VSB modulator.

hold during the transition interval $|f| \in [f_c - f_v, f_c + f_v]$:

(a) The sum of the values of $|H(f)|$ at any two frequencies equally displaced above and below f_c is fixed (normalized to unity such that $|H(f_c)| = 1/2$).
(b) The phase response, $\arg(H(f))$, is linear. That is,

$$H(f - f_c) + H(f + f_c) = 1 \quad \text{for } |f| \le W. \qquad (3.11)$$

Consequently, denoting the message bandwidth by W and the width of the VSB by f_v, the transmission bandwidth of VSB modulation is given by

$$B_T = W + f_v. \qquad (3.12)$$

The VSB modulated waveform is thus expressed us

$$s(t) = \frac{A_c}{2} m(t) \cos(2\pi f_c t) \pm \frac{A_c}{2} m^Q(t) \sin(2\pi f_c t), \qquad (3.13)$$

where the plus/minus sign corresponds to the transmission of a vestige of upper/lower sideband. Signal $m^Q(t)$ in the quadrature component is obtained by passing message signal $m(t)$ through a filter with frequency response

$$H_Q(f) = j[H(f - f_c) - H(f + f_c)] \quad \text{for } |f| \le W. \qquad (3.14)$$

VSB has been widely applied in TV broadcasting, not only for analog TV but also for digital TV (i.e., ATSC system in North America and Korea), due to the following reasons:

- The TV signal (i.e., video signal) has a large bandwidth and a good amount of low-frequency content, while VSB

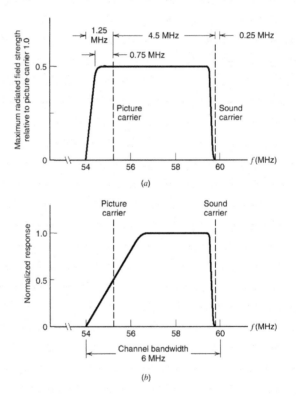

Fig. 3.14 VSB for analog TV.

modulation fits this feature. Figure 3.14 shows a VSB signal bandwidth for analog TV.

- We may use envelope detection for VSB-modulated waveform with an additional carrier, which results in simple design of the receiver.

3.4 Frequency Division Multiplexing (FDM)

We may note that the product modulator (i.e., a sort of mixer) has an interesting function to translate signal to higher frequency. Such a function block to form *frequency translation* as shown in Figure 3.15 can be called an analog waveform *mixer*. Recall that we must include a BPF after mixing the waveform.

Also recall that mixing a sinusoidal waveform is equivalent to translating an original waveform into positive and negative frequency directions at the same time. Consequently, the mixer can execute up-conversion or down-conversion in frequency translation. Frequency translation is a linear operation of waveforms.

Once we can execute frequency translation, we can facilitate frequency multiplexing as one of the purposes of modulation. We can translate different signals into different portions of frequency spectrum, so that signals are separated in frequency domain, which is known as FDM. Figure 3.16 depicts the block diagram of an FDM system, which enables our handling of multiple analog signals. Its counterpart to distinguish in time domain is called *time division multiplexing* and is widely used in digital communication systems and networks.

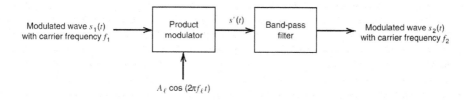

Fig. 3.15 Mixer.

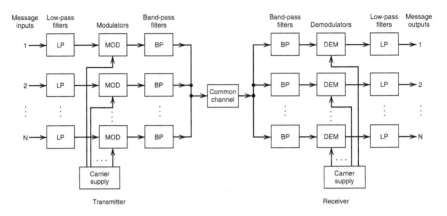

Fig. 3.16 FDM system.

3.5   Angle Modulation

Let $\theta_i(t)$ denote the angle of a modulated sinusoidal carrier. The resulting angle-modulated waveform is expressed as

$$s(t) = A_c \cos[\theta_i(t)], \tag{3.15}$$

where A_c is the carrier amplitude. If $\theta_i(t)$ increases monotonically with time over $[t, t + \Delta t]$, then

$$f_{\Delta t}(t) = \frac{\theta_i(t + \Delta t) - \theta_i(t)}{2\pi \Delta t}. \tag{3.16}$$

We may therefore define the *instantaneous frequency* of $s(t)$ as

$$f_i(t) = \lim_{\Delta t \to 0} f_{\Delta t}(t)$$

$$= \frac{1}{2\pi} \frac{d\theta_i(t)}{dt}. \tag{3.17}$$

For an un-modulated carrier, the angle $\theta_i(t) = 2\pi f_c t + \phi_c$ and thus the angle-modulated signal $s(t)$ is a rotating phasor of amplitude A_c and angle $\theta_i(t)$. On the basis of this interpretation, we may define two common modulations as follows.

Definition 3.1. *Phase modulation* (PM) is a form of angle modulation where the angle $\theta_i(t)$ is linearly varied with the message signal $m(t)$, whose signal time-domain representation is

$$s(t) = A_c \cos\left[2\pi f_c t + k_p m(t)\right],\qquad(3.18)$$

where k_p is a constant representing the *phase* *sensitivity* of the modulation.

Definition 3.2. *Frequency modulation* (FM) is a form of angle modulation where the instantaneous frequency $f_i(t)$ is linearly varied with the message signal $m(t)$, i.e., $f_i(t) = f_c + k_f m(t)$, whose time-domain representation is

$$s(t) = A_c \cos\left[2\pi f_c t + 2\pi k_f \int_0^t m(x)dx\right],\qquad(3.19)$$

where k_f is a constant representing the *frequency* *sensitivity* of the modulation.

Figure 3.17 illustrates the relationship between FM and PM.

3.6 Frequency Modulation

Equation (3.92) defines FM modulation, which is the result of integration, and thus FM is a nonlinear modulation. We may further observe that

- Message is carried with integration in FM, which is harder to attenuate.
- Message is stored in frequency (e.g., information embedded into zero crossings), which is much more robust against channel noise and impairments than AM.

Fig. 3.17 (a) Generating FM through PM; (b) generating PM through FM.

For ease of spectral analysis later, we consider a sinusoidal modulating signal defined as

$$m(t) = A_m \cos(2\pi f_m t). \tag{3.20}$$

The instantaneous frequency of the resulting FM signals is given by

$$f_i(t) = f_c + k_f A_m \cos(2\pi f_m t)$$
$$= f_c + \Delta f \cos(2\pi f_m t), \tag{3.21}$$

where Δf is the *frequency deviation* and is given by

$$\Delta f = k_f A_m. \tag{3.22}$$

The angle $\theta_i(t)$ of the FM signal is expressed as

$$\theta_i(t) = 2\pi \int_0^t f_i(\tau)\, d\tau$$
$$= 2\pi f_c t + \frac{\Delta f}{f_m} \sin(2\pi f_m t). \tag{3.23}$$

We usually define the modulation index of the FM signal as the ratio of frequency deviation Δf to modulation frequency f_m, which is expressed as

$$\beta = \frac{\Delta f}{f_m} \tag{3.24}$$

and

$$\theta_i(t) = 2\pi f_c t + \beta \sin(2\pi f_m t). \tag{3.25}$$

Then, the FM signal is given by

$$s(t) = A_c \cos[2\pi f_c t + \beta \sin(2\pi f_m t)]. \tag{3.26}$$

Depending on the value of modulation index β, we usually have two categories of FM:

- Narrowband FM for $\beta \ll 1$.
- Wideband FM, $\beta \gg 1$.

3.6.1 Narrowband FM

We may expand (3.26) as

$$s(t) = A_c \cos(2\pi f_c t) \cos[\beta \sin(2\pi f_m t)] - A_c \sin(2\pi f_c t) \sin[\beta \sin(2\pi f_m t)].$$
(3.27)

Assuming β is small (i.e., $\beta \ll 1$), we can approximate

$$\cos[\beta \sin(2\pi f_m t)] \cong 1$$
$$\sin[\beta \sin(2\pi f_m t)] \cong \beta \sin(2\pi f_m t).$$
(3.28)

Hence,

$$s(t) \cong A_c \cos(2\pi f_c t) - \beta A_c \sin(2\pi f_c t) \sin(2\pi f_m t).$$
(3.29)

(3.29) approximates a narrowband FM signal generated by a sinusoidal modulating signal (3.20). We can derive a modulator as shown in Figure 3.18. An ideal FM signal shall have a constant envelope. For a sinusoidal modulating signal of frequency f_m, the angle $\theta_i(t)$ is also sinusoidal with the same frequency. However, the modulated signal produced as shown in Figure 3.18 is different from the ideal condition in two aspects:

(a) The envelope contains a residual AM and thus varies with time.
(b) For a sinusoidal modulating wave, the angle $\theta_i(t)$ contains harmonic distortion in higher orders.

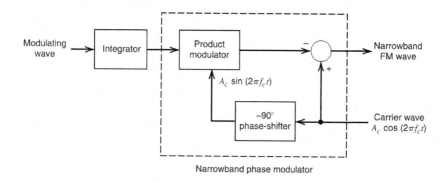

Narrowband phase modulator

Fig. 3.18 Block diagram to generate a narrowband FM.

By restricting the modulation index β to be ≤ 0.3, the effects of residual AM and harmonic PM are limited to negligible levels.

We may further expand (3.29) as

$$s(t) \cong A_c \cos(2\pi f_c t) + \frac{1}{2}\beta A_c \{\cos[2\pi(f_c + f_m)t] - \cos[2\pi(f_c - f_m)t]\}.$$
(3.30)

This is similar to the corresponding AM signal

$$s_{\mathrm{AM}}(t) \cong A_c \cos(2\pi f_c t) + \frac{1}{2}\mu A_c \{\cos[2\pi(f_c + f_m)t]$$
$$+ \cos[2\pi(f_c - f_m)t]\},$$
(3.31)

where μ is the modulation factor of the AM signal. Comparing the above two equations, we may observe the difference between AM and narrowband FM to be the sign of lower sideband. It suggests the same transmission bandwidth (i.e., $2f_m$) for these two systems. Their phasor diagrams are shown in Figure 3.19.

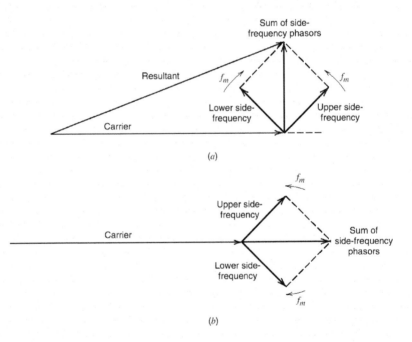

Fig. 3.19 Phasor diagram for sinusoidal modulations: (a) narrowband FM; (b) AM.

3.6.2   Wideband FM

Assuming that the carrier frequency f_c is larger enough (relative to the bandwidth of an FM signal), (3.26) can be expressed as

$$s(t) = \text{Re}[A_c e^{j2\pi f_c t + j\beta \sin(2\pi f_m t)}]$$
$$= \text{Re}[\tilde{s}(t) e^{j2\pi f_c t}], \tag{3.32}$$

where $\tilde{s}(t)$ is the complex envelope of FM signal $s(t)$ and is given by

$$\tilde{s}(t) = A_c e^{j\beta \sin(2\pi f_m t)}. \tag{3.33}$$

We may further expand $\tilde{s}(t)$ as a complex Fourier series

$$\tilde{s}(t) = \sum_{n=-\infty}^{\infty} c_n e^{j2\pi n f_m t}. \tag{3.34}$$

where the coefficients are given by

$$c_n = f_m \int_{-1/2f_m}^{1/2f_m} \tilde{s}(t) e^{-j2\pi n f_m t} dt$$
$$= f_m A_c \int_{-1/2f_m}^{1/2f_m} e^{j\beta \sin(2\pi f_m t) - j2\pi n f_m t} dt. \tag{3.35}$$

Defining a new variable $x = 2\pi f_m t$, (3.35) is in the form

$$c_n = \frac{A_c}{2\pi} \int_{-\pi}^{\pi} e^{j(\beta \sin x - nx)} dx. \tag{3.36}$$

The nth-order Bessel function of the first kind with argument β is denoted as $J_n(\beta)$, which is given by

$$J_n(\beta) = \frac{1}{2\pi} \int_{-\pi}^{\pi} e^{j(\beta \sin x - nx)} dx. \tag{3.37}$$

The complex envelope of the FM signal is expressed as

$$\tilde{s}(t) = A_c \sum_{n=-\infty}^{\infty} J_n(\beta) e^{j2\pi n f_m t}. \tag{3.38}$$

Then, (3.32) becomes

$$s(t) = A_c \cdot \text{Re}\left[\sum_{n=-\infty}^{\infty} J_n(\beta) e^{j2\pi n f_m t} \right]$$

$$= A_c \sum_{n=-\infty}^{\infty} J_n(\beta) \cos[2\pi(f_c + n f_m)t]. \tag{3.39}$$

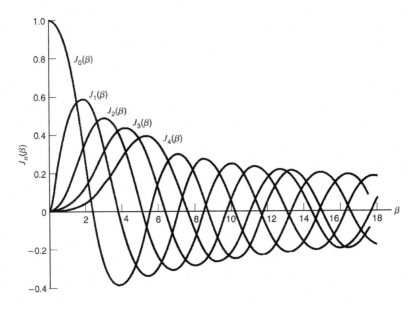

Fig. 3.20 Bessel functions of the first kind.

The discrete spectrum of $s(t)$ is obtained by using a Fourier transform:

$$S(f) = \frac{A_c}{2} \sum_{n=-\infty}^{\infty} J_n(\beta)\left[\delta(f - f_c - nf_m) + \delta(f + f_c + nf_m)\right]. \quad (3.40)$$

The Bessel function $J_n(\beta)$ is plotted as shown in Figure 3.20 and its property is as follows.

Property 3.3 (Bessel function $J_n(\beta)$).

(a) $J_n(\beta) = (-1)^n J_{-n}(\beta)$ for all n $\qquad\qquad$ (3.41)

(b) For $\beta \ll 1$,

$$J_0(\beta) \cong 1$$
$$J_1(\beta) \cong \frac{\beta}{2} \qquad\qquad (3.42)$$
$$J_n(\beta) \cong 0, \quad n > 2.$$

(c) $\displaystyle\sum_{n=-\infty}^{\infty} J_n^2(\beta) = 1.$ $\qquad\qquad\qquad\qquad$ (3.43)

For an FM spectrum with sinusoidal signals, we may reach the following observations based on the above explorations of Bessel functions.

- The FM spectrum consists of a carrier and an infinite number of side-lobe frequency components distributed symmetrically on each side of carrier, at frequency separations of $f_m, 2f_m, 3f_m, \ldots$.
- For small β, only the Bessel coefficients $J_0(\beta)$ and $J_1(\beta)$ have non-trivial values. Therefore, the FM signal is effectively composed of a carrier and a pair of frequency components $f_c \pm f_m$, which is equivalent to a special case of narrowband FM.
- The amplitude of the carrier varies with β following $J_0(\beta)$, which suggests that the amplitude of a carrier in FM depends on the modulation index β. From (3.39), the average power of an FM signal is given by

$$P = \frac{A_c^2}{2} \sum_{n=-\infty}^{\infty} J_n^2(\beta) = \frac{A_c^2}{2}. \tag{3.44}$$

Theoretically, an FM signal consists of an infinite number of side frequencies so that the bandwidth is infinite. Practically, the FM signal can be limited to a finite number of significant side frequencies with a specified distortion, which allows us to specify an effective bandwidth for the transmission of an FM signal. We may observe this from Figure 3.21, where Δf means the frequency deviation of FM.

We first consider an FM signal generated by a single-tone modulating wave of frequency f_m. Continuing from the above observations, for small β, the FM signal bandwidth is therefore approaching $2f_m$.

Proposition 3.1 (Carson's Rule). An empirical approximation for the transmission bandwidth of an FM signal generated by a single-tone modulating wave of frequency f_m is known as *Carson's rule*, which is expressed as

$$B_T \cong 2\Delta f + 2f_m = 2\Delta f \left(1 + \frac{1}{\beta}\right). \tag{3.45}$$

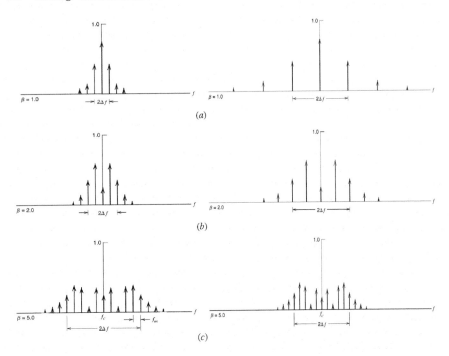

Fig. 3.21 FM spectrum for sinusoidal modulation (a) of fixed frequency and varying amplitude (b) of varying frequency and fixed amplitude.

We may also consider the bandwidth as $2n_{\max}f_m$, while $n_{\max}$ is the largest integer satisfying $|J_n(\beta)| > 0.01$. For a more general case of arbitrary modulating signal $m(t)$ with the highest frequency component at W, the required transmission bandwidth is estimated by the worst case. The *deviation ratio* $D = \frac{\Delta f_{\max}}{W}$ can be used with Carson's rule to determine FM transmission bandwidth.

There are two fundamental structures to generate FM signals:

- Direct FM: The carrier frequency is directly varied according to the baseband signal by using a VCO.
- Indirect FM: The signal is first used to generate a narrowband FM signal, and then frequency multiplication is employed to increase frequency deviation up to a desired level. Figure 3.22 depicts this structure to tolerate more oscillator instability.

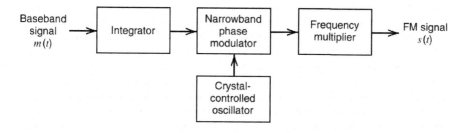

Fig. 3.22 Block diagram of an indirect wideband FM modulation.

The frequency multiplier can be realized by passing a signal to a nonlinear device with an input–output transfer function as $H(x) = a_0 + a_1x + a_2x^2 + a_3x^3 + \cdots$, where x is the input and $H(x)$ is the output. From this transfer function, we observe the generation of high-order harmonics. Then, we use an appropriate BPF with selected mid-band frequency to filter out what we want. This concept is useful in many aspects in communication systems to generate high-order harmonics or to understand nonlinear generation of undesired harmonics.

Demodulation of FM signals can be facilitated via direct method using *frequency discriminator*, or via indirect method using PLL. Frequency discriminator consists of a slope circuit followed by an envelope detector. The ideal slope circuit has a purely imaginary frequency response and is varying linearly with frequency inside a prescribed frequency interval, which is given by

$$H_1(f) = \begin{cases} j2\pi a\left(f - f_c + \frac{B_T}{2}\right) & |f - f_c| \leq \frac{B_T}{2} \\ j2\pi a\left(f + f_c - \frac{B_T}{2}\right) & |f + f_c| \leq \frac{B_T}{2} \\ 0 & \text{elsewhere.} \end{cases} \tag{3.46}$$

We may replace the BPF with frequency response $H_1(f)$ by an equivalent LPF with frequency response $\tilde{H}_1(f)$ under the following conditions:

(a) $\tilde{H}_1(f - f_c)$;
(b) $\tilde{H}_1(f - f_c) = 2H_1(f)$ for $f > 0$.

Then,

$$\tilde{H}_1(f) = \begin{cases} j4\pi a\left(f + \frac{B_T}{2}\right) & |f| \le \frac{B_T}{2} \\ 0 & \text{elsewhere.} \end{cases} \tag{3.47}$$

The complex envelope of $s(t)$ is given by

$$\tilde{s}(t) = A_c e^{j2\pi k_f \int_0^t m(\tau)d\tau}. \tag{3.48}$$

The Fourier transform of $\tilde{s}_1(t)$, $\tilde{S}_1(f)$, is given by

$$\tilde{S}_1(f) = \frac{1}{2}\tilde{H}_1(f)\tilde{S}(f)$$

$$= \begin{cases} j2\pi a\left(f + \frac{B_T}{2}\right)\tilde{S}(f) & |f| \le \frac{B_T}{2} \\ 0 & \text{elsewhere.} \end{cases} \tag{3.49}$$

By using the differentiation property of Fourier transform,

$$\tilde{s}_1(t) = a\left[\frac{d\tilde{s}(t)}{dt} + j\pi B_T \tilde{s}(t)\right]$$

$$= j\pi B_T a A_c \left[1 + \frac{2k_f}{B_T}m(t)\right] e^{j2\pi k_f \int_0^t m(\tau)d\tau}. \tag{3.50}$$

The desired response of the slope circuit is given by

$$s_1(t) = \text{Re}\left[\tilde{s}_1(t)e^{j2\pi f_c t}\right]$$

$$= \pi B_T a A_c \left[1 + \frac{2k_f}{B_T}m(t)\right] \cos\left[2\pi f_c t + 2\pi k_f \int_0^t m(\tau)d\tau + \frac{\pi}{2}\right]. \tag{3.51}$$

Suppose that we select

$$\left|\frac{2k_f}{B_T}m(t)\right| < 1 \quad \forall t.$$

The resulting envelope detector output is therefore expressed as

$$|\tilde{s}_1(t)| = \pi B_T a A_c \left[1 + \frac{2k_f}{B_T}m(t)\right]. \tag{3.52}$$

To remove the bias term in the above equation and to obtain the slope, we may design a complimentary slope circuit with frequency response $H_2(f)$. These two slope circuits are related by

$$\tilde{H}_2(f) = \tilde{H}_1(-f). \tag{3.53}$$

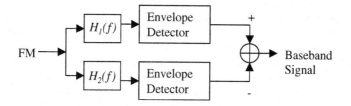

Fig. 3.23 Balanced frequency discriminator.

Similarly,

$$|\tilde{s}_2(t)| = \pi B_T a A_c \left[1 - \frac{2k_f}{B_T} m(t)\right]. \tag{3.54}$$

The difference between the two envelopes is expressed as

$$s_o(t) = |\tilde{s}_1(t)| - |\tilde{s}_2(t)| = 4\pi k_f a A_c m(t). \tag{3.55}$$

Consequently, we can obtain the original message signal with a proportional constant, and free of bias term. Such an ideal frequency discriminator as shown in Figure 3.23 is called a *balanced frequency discriminator*.

The next step toward better voice/audio communication is the *stereo multiplexing*, which is in fact a form of FDM to transmit two separate signals (left and right) over the same carrier. The specifications of stereo FM have two constraints:

- The transmission bandwidth must be within the originally allocated FM broadcast channels.
- It must be backward compatible with legacy monophonic radio receivers.

Let $m_l(t)$ and $m_r(t)$ be the signals picked up from microphones at the right-hand side and left-hand side, respectively. From them, we form the *sum signal* $m_l(t) + m_r(t)$ and the *difference signal* $m_l(t) - m_r(t)$, while the sum signal is still good for monophonic reception. Figure 3.24 depicts the stereo FM systems. If we define a pilot of $f_c = 19\,\text{kHz}$ as shown in Figure 3.24 (pilot here as a reference for the purpose of coherent detection) and K as the pilot amplitude, the multiplexed

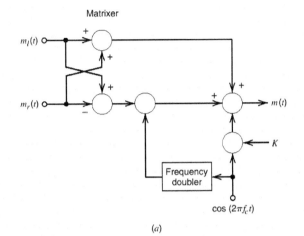

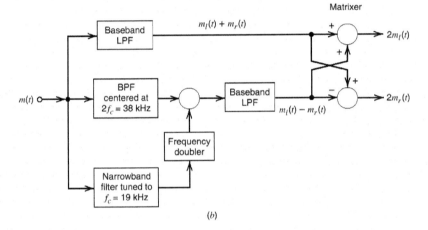

Fig. 3.24 FM stereo: (a) multiplexer at the transmitter; (b) de-multiplexer at the receiver.

signal is thus given by

$$m(t) = [m_l(t) + m_r(t)] + [m_l(t) - m_r(t)]\cos(4\pi f_c t) + K\cos(2\pi f_c t). \tag{3.56}$$

The pilot is usually allocated 8%–10% of peak frequency deviation. De-multiplexing is generally a direct reverse operation of multiplexing.

Recall the nonlinear generation of harmonics from earlier discussions. Nonlinear effects can be very "influential" in communication systems. In what follows, we will examine the impacts of weak nonlinearity on FM. Let us consider the IO relationship as follows:

$$v_o(t) = a_1 v_i(t) + a_2 v_i^2(t) + a_3 v_i^3(t). \tag{3.57}$$

Equation (2.73) can represent a memoryless "channel" due to being an instantaneous function of inputs. For an FM signal as the input, we have

$$v_i(t) = A_c \cos[2\pi f_c t + \phi(t)], \quad \text{where } \phi(t) = 2\pi k_f \int_0^t m(\tau) \, d\tau.$$

Then, the output is given by

$$v_o(t) = a_1 A_c \cos[2\pi f_c t + \phi(t)] + a_2 A_c^2 \cos^2[2\pi f_c t + \phi(t)]$$
$$+ a_3 A_c^3 \cos^3[2\pi f_c t + \phi(t)]. \tag{3.58}$$

After some algebras, we have

$$v_o(t) = \frac{1}{2} a_2 A_c^2 + \left(a_1 A_c + \frac{3}{4} a_3 A_c^3 \right) \cos[2\pi f_c t + \phi(t)]$$
$$+ \frac{1}{2} a_2 A_c^2 \cos[4\pi f_c t + 2\phi(t)] + \frac{1}{4} a_3 A_c^3 \cos[6\pi f_c t + 3\phi(t)]. \tag{3.59}$$

We may observe linear, second-order, and third-order terms in the above equation. Again, Δf denotes the frequency deviation of $v_i(t)$, and W denotes the highest frequency component of the message $m(t)$. Applying Carson's rule, the necessary condition to separate an FM signal and the carrier is

$$2f_c - (2\Delta f + W) > f_c + \Delta f + W \quad \text{or} \quad f_c > 3\Delta f + 2W. \tag{3.60}$$

By applying a BPF of mid-band frequency f_c and bandwidth $2\Delta f + 2W$, the channel output is therefore given by

$$v_o'(t) = \left(a_1 A_c + \frac{3}{4} a_3 A_c^3 \right) \cos[2\pi f_c t + \phi(t)]. \tag{3.61}$$

Surprisingly, through appropriate filtering, the only impact of nonlinearity on FM is the modification of signal amplitude that has less effect on the correct reception of the FM signal. It suggests advantages for FM in nonlinear channels, such as those using very nonlinear power amplifier, etc. FM is supposed to be more sensitive to phase nonlinearity. There is a well-known situation, *AM–PM conversion*, which plays an important role in system performance. For highly nonlinear channels such as satellite communications, AM–PM conversion is critical.

3.7   Superhetrodyne Receiver

A receiver should demodulate the signals. However, there are some other required functions to ensure the success of demodulation:

- Carrier frequency tuning, so that desired signal can be selected.
- Filtering, to separate the desired signal from other modulated signals or harmonics.
- Amplification, to compensate the loss of signal power/ strength after transmission through the channel.

Receiver structure is thus very much needed to design a communication system. *Superheterodyne receiver* has been used since early days in communications and broadcasting Figure 3.25. The major feature of a superheterodyne receiver is to use a pre-determined *intermediate frequency* (IF), usually lower than the carrier frequency. The heterodyne process produces an IF (f_{IF}) carrier and satisfies

$$f_{IF} = f_{LO} - f_c, \tag{3.62}$$

where f_{LO} is the frequency generated by the local oscillator.

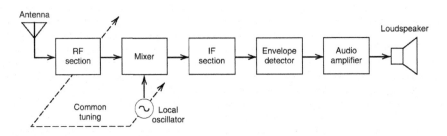

Fig. 3.25 Block diagram of an AM superheterodyne receiver.

In contrast, in modern wireless communication and broadcasting, to deliver a single system-on-chip (SoC), we may commonly use a *direct-conversion* (DC) receiver structure, without an IF section. DC receiver directly converts a radio frequency signal waveform into baseband processing to extract a message. Such an approach might also be known as zero-IF in wireless CMOS SoC, compared with low-IF (IF close to but higher than the highest frequency component of a baseband signal). The most advanced structure is to develop a digital radio processor to significantly reduce RF section, and thus suitable for deep-submicron CMOS wireless SoC.

3.8 Performance Analysis in Noise

It is always important in system engineering (such as communication engineering) to "foresee" the system performance prior to implementation, especially in modern systems with great complexity. From now on, we explore the system performance of analog communication through noisy channels, which is closer to practical engineering world. We start from considering channel noise into performance analysis, which involves

- Channel: We assume the communication channel to be distortion-less, and the signal is perturbed only by the additive white Gaussian noise (AWGN) defined in Chapter 2.
- Ideal filtering: The receiver consists of an ideal band-pass filtering followed by a perfectly appropriate demodulator.

In analog communications and even in most digital communications, the *signal-to-noise ratio* (*SNR*, in the unit of dB, actually unit-less) is considered as a reliable performance index, which is given by

$$SNR = \frac{\text{signal power}}{\text{noise power}}. \tag{3.63}$$

For AWGN with two-side power spectral density (psd) $\frac{N_0}{2}$, its power spectrum after an ideal low-pass filtering with stop-frequency W is as shown in Figure 3.26, whose total noise power is $P_{\text{total−noise}} = \frac{N_0}{2}(2W) = WN_0$.

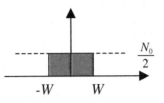

Fig. 3.26 AWGN spectrum after LPF.

Definition 3.4. For the demodulated signal and filtered noise, we can define

(a) SNR_I, input SNR, as the ratio of the average power of the modulated signal to the average power of the filtered noise.

(b) SNR_O, output SNR, as the ratio of the average power of the demodulated message signal to the average power of the noise, both measured at the receiver output.

(c) SNR_C, channel SNR, as the ratio of the average power of the modulated signal to the average power of channel noise in the message bandwidth, both measured at the receiver input.

Definition 3.5. We define the *figure of merit* for the receiver as

$$\text{Figure of merit} = \frac{SNR_O}{SNR_C}.$$

Since usually $SNR_C > SNR_O$, the maximum figure of merit $= 1$ for receivers.

It is also interesting to note that processing cannot increase the figure of merit, which is consistent with the signal processing theorem in information theory. It is sufficient to successfully demodulate the message signals, in system design, since the SNR is the performance index for analog modulation and we cannot increase the SNR by processing. Note that the figure of merit is not constrained to 1 for some components.

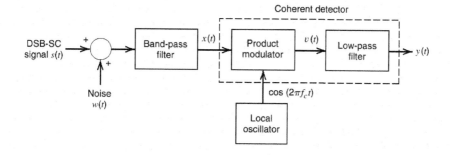

Fig. 3.27 Block diagram of a coherent DSB-SC receiver.

3.8.1 Coherent Detection in an AM Receiver

Demodulation of an AM signal depends on the existence of a carrier or not. If the carrier is suppressed, we usually use a linear coherent receiver. If the AM transmission includes a carrier, we can simply use an envelope detector for demodulation, which involves a nonlinear receiver structure. We start from coherent detection of an AM signal with a receiver model as shown in Figure 3.27. Band-pass filtered waveform $x(t)$ is mixed with a locally generated sinusoidal wave, and then low-pass filtered to obtain a demodulated waveform $y(t)$. This BPF is also known as a receiving filter.

The DSB-SC component of the filtered signal $x(t)$ is represented as

$$s(t) = CA_c \cos(2\pi f_c t)m(t). \qquad (3.64)$$

Here, we introduce a system-dependent scaling factor C to align the measurement of signal and noise. We assume that message $m(t)$ is from a stationary process with zero mean and psd $S_M(f)$ limited to a maximum frequency W (message bandwidth). The average power of message signal is given by

$$P = \int_{-W}^{W} S_M(f)\,df. \qquad (3.65)$$

The average power of a DSB-SC modulated signal is $C^2 A_c^2 P/2$. With two-side noise psd and message bandwidth W, the channel SNR of DSB-SC is given by

$$(SNR)_{\text{C,DSB}} = \frac{C^2 A_c^2 P/2}{W N_0}. \qquad (3.66)$$

To determine the out SNR, using the narrowband representation of the filtered noise $n(t)$, the total signal waveform at the input of the coherent detector is expressed as

$$x(t) = s(t) + n(t)$$

$$= CA_c \cos(2\pi f_c t)m(t) + n_I(t)\cos(2\pi f_c t) - n_Q(t)\sin(2\pi f_c t). \tag{3.67}$$

The output of a product modulator component at the coherent detector can be obtained as

$$v(t) = x(t)\cos(2\pi f_c t)$$

$$= \frac{1}{2}CA_c m(t) + \frac{1}{2}n_I(t) + \frac{1}{2}[CA_c m(t) + n_I(t)]\cos(4\pi f_c t)$$

$$- \frac{1}{2}n_Q(t)\sin(4\pi f_c t). \tag{3.68}$$

After low-pass filtering to remove HF terms, the receiver output is given by

$$y(t) = \frac{1}{2}CA_c m(t) + \frac{1}{2}n_I(t). \tag{3.69}$$

We may observe that only message signal plus low-pass equivalent in-phase noise appears at the receiver output, while quadrature noise component is rejected. From (3.69), the average message signal power is $C^2 A_c^2 P/4$, and the average noise at the receiver output is $(\frac{1}{2})^2 2W N_0 = \frac{1}{2}W N_0$. We consequently obtain the out SNR for DSB-SC using coherent detection as

$$(\text{SNR})_{O,\text{DSB-SC}} = \frac{C^2 A_c^2 P/4}{W N_0/2}$$

$$= \frac{C^2 A_c^2 P}{2W N_0}. \tag{3.70}$$

Then, the figure of merit is given by

$$\left.\frac{(SNR)_O}{(SNR)_C}\right|_{\text{DSB-SC}} = 1. \tag{3.71}$$

In a similar manner (see exercise), a coherent SSB receiver has exactly the same output SNR as that of the DSB-SC receiver, and also the same figure of merit.

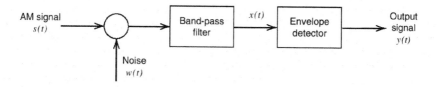

Fig. 3.28 AM envelope detection.

3.8.2 Noise in AM with Envelope Detection

The envelope detection of an AM signal under AWGN is represented in Figure 3.28. An AM signal with double sidebands and a carrier is represented as

$$s(t) = A_c[1 + k_a m(t)]\cos(2\pi f_c t). \tag{3.72}$$

The average power of a carrier component is $A_c^2/2$, and the average power of a message-bearing component is $A_c^2 k_a^2 P/2$, while again the average power of a message is P. The noise power is $W N_0$. The channel *SNR* is therefore given by

$$(SNR)_{C,AM} = \frac{A_c^2(1 + k_a^2 P)/2}{W N_0}. \tag{3.73}$$

To evaluate the output SNR, the filtered output $x(t)$ applied to the envelope detector is given by

$$
\begin{aligned}
x(t) &= s(t) + n(t) \\
&= [A_c + A_c k_a m(t) + n_I(t)]\cos(2\pi f_c t) - n_Q(t)\sin(2\pi f_c t). \tag{3.74}
\end{aligned}
$$

It is useful to consider via phasors as shown in Figure 3.29. The receiver output is obtained as

$$
\begin{aligned}
y(t) &= \text{envelope of } x(t) \\
&= \{[A_c + A_c k_a m(t) + n_I(t)]^2 + n_Q^2(t)\}^{1/2}. \tag{3.75}
\end{aligned}
$$

In usual cases, to operate normally, the average carrier power is greater than the average noise power, which implies that the carrier term is usually greater than the noise term. We may approximate

$$y(t) \cong A_c + A_c k_a m(t) + n_I(t). \tag{3.76}$$

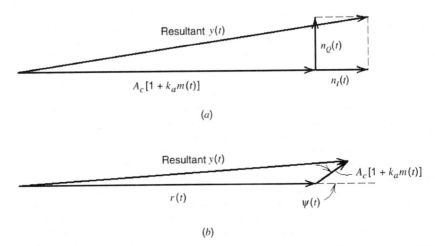

Fig. 3.29 Phasor diagram for AM plus narrowband noise: (a) high carrier-to-noise ratio; (b) low carrier-to-noise ratio.

The DC term, A_c, does not carry any information. The second term is useful toward average signal power. The output *SNR* at the AM envelope detector is approximately

$$(SNR)_{O,AM} \cong \frac{A_c^2 k_a^2 P}{2W N_0}. \tag{3.77}$$

Under the conditions of small noise power (relative to average carrier power) and $k_a < 1$, the figure of merit for AM is given by

$$\left.\frac{(SNR)_O}{(SNR)_C}\right|_{AM} \cong \frac{k_a^2 P}{1 + k_a^2 P}. \tag{3.78}$$

We may observe that the figure of merit in (3.78) $\frac{k_a^2 \rho}{1+k_a^2 \rho} < 1$, whereas the figure of merit for DSB-SC is equal to 1. It suggests that AM envelope detection is always inferior to that of DSB-SC coherent detection. In fact, non-coherent detection (such as envelope detection) is inferior to coherent detection, for signaling methods commonly used in communications.

When the carrier-to-noise ratio (*CNR*) is small (relative to unity), the noise term dominates the performance evaluation. It results in a loss of message at the envelope detector at low *CNR*, and it is known as *threshold effect*. Intuitively, when the *CNR* drops, the signal may quickly get lost, which we call threshold effect.

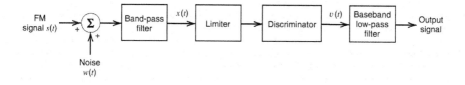

Fig. 3.30 Block diagram of an FM receiver.

3.8.3 Noise in FM Receivers

Our performance analysis for FM in noise starts from Figure 3.30. The noise $w(t)$ is AWGN with zero mean and (two-side) psd $N_0/2$. The received FM signal $s(t)$ has a carrier frequency f_c and transmission bandwidth B_T, such that the power outside the frequency band $[f_c - B_T/2, f_c + B_T/2]$ is negligible. The BPF has a mid-band frequency as the carrier and bandwidth $B_T \ll f_c$. Narrowband noise representation can be therefore employed. The limiter (or clipper) is to remove amplitude variation. Frequency discriminator is just as the earlier description in FM. Baseband LPF is also known as *post-detection filter* to accommodate the message signal.

The filtered noise as narrowband noise canonical representation is

$$n(t) = n_I(t)\cos(2\pi f_c t) - n_Q(t)\sin(2\pi f_c t).$$

An equivalent representation in envelope and phase is

$$n(t) = r(t)\cos[(2\pi f_c t) + \psi(t)], \qquad (3.79)$$

where the envelope is expressed as

$$r(t) = [n_I^2(t) + n_Q^2(t)]^{1/2} \qquad (3.80)$$

and the phase is expressed as

$$\psi(t) = \tan^{-1}\left[\frac{n_Q(t)}{n_I(t)}\right]. \qquad (3.81)$$

We may recall that the envelope is Rayleigh distributed and the phase is uniformly distributed. The FM signal is given by

$$s(t) = A_c \cos\left[2\pi f_c t + 2\pi k_f \int_0^t m(\tau)d\tau\right]. \qquad (3.82)$$

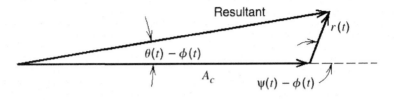

Fig. 3.31 Phasor diagram of FM with high *CNR*.

and we define

$$\phi(t) = 2\pi k_f \int_0^t m(\tau)\,d\tau. \tag{3.83}$$

Therefore,

$$s(t) = A_c \cos[2\pi f_c t + \phi(t)]. \tag{3.84}$$

The noisy signal at the BPF output is given by

$$x(t) = s(t) + n(t)$$
$$= A_c \cos[2\pi f_c t + \phi(t)] + r(t)\cos[(2\pi f_c t) + \psi(t)]. \tag{3.85}$$

Again, we apply the phasor diagram as shown in Figure 3.31. The resulting phase $\theta(t)$ representing $x(t)$ is obtained as

$$\theta(t) = \phi(t) + \tan^{-1}\left\{\frac{r(t)\sin[\psi(t) - \phi(t)]}{A_c + r(t)\cos[\psi(t) - \phi(t)]}\right\}. \tag{3.86}$$

We have little interest in the envelope of $x(t)$ because envelope variations can be removed by the limiter. We should focus on the errors in the instantaneous frequency caused by the noise. For high *CNR*, we can approximate

$$\theta(t) \cong \phi(t) + \frac{r(t)}{A_c}\sin[\psi(t) - \phi(t)]. \tag{3.87}$$

Using (3.83),

$$\theta(t) \cong 2\pi k_f \int_0^t m(\tau)\,d\tau + \frac{r(t)}{A_c}\sin[\psi(t) - \phi(t)]. \tag{3.88}$$

The discriminator output is given by

$$v(t) = \frac{1}{2\pi}\frac{d\theta(t)}{dt} = k_f m(t) + n_d(t) \tag{3.89}$$

and is proportional to the message, where the noise is given by

$$n_d(t) = \frac{1}{2\pi A_c} \frac{d}{dt}\{r(t)\sin[\psi(t) - \phi(t)]\}. \tag{3.90}$$

Since the phase $\psi(t)$ is uniformly distributed over $[0, 2\pi]$, we may assume that $\psi(t) - \phi(t)$ is also uniformly distributed over $[0, 2\pi]$ and thus the noise $n_d(t)$ at the discriminator output is independent of the modulating signal. It is true for high CNR, and we can obtain

$$n_d(t) \cong \frac{1}{2\pi A_c} \frac{d}{dt}\{r(t)\sin[\psi(t)]\}. \tag{3.91}$$

Note that

$$n_Q(t) = r(t)\sin[\psi(t)]. \tag{3.92}$$

We can re-write it as

$$n_d(t) \cong \frac{1}{2\pi A_c} \frac{dn_Q(t)}{dt}. \tag{3.93}$$

It suggests that the noise $n_d(t)$ at the discriminator output is determined effectively by the carrier amplitude A_c and the quadrature component $n_Q(t)$ of narrowband noise $n(t)$.

From (3.89), the average output signal power is $k_f^2 P$. By the differentiation property of Fourier transform,

$$\frac{j2\pi f}{2\pi A_c} = \frac{jf}{A_c}$$

the power spectral density $S_{N_d}(f)$ of $n_d(t)$ is related to the psd of $n_Q(t)$, $S_{N_Q}(f)$, as

$$S_{N_d}(f) = \frac{f^2}{A_c^2} S_{N_Q}(f). \tag{3.94}$$

Since $S_{N_Q}(f)$ is from the low-pass filtered AWGN, we have

$$S_{N_d}(f) = \begin{cases} \dfrac{N_0 f^2}{A_c^2}, & |f| \leq \frac{B_T}{2} \\ 0, & \text{otherwise.} \end{cases} \tag{3.95}$$

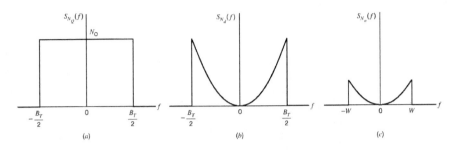

Fig. 3.32 Noise analysis of FM.

For wideband FM, W is usually less than B_T, we can further reach

$$S_{N_O}(f) = \begin{cases} \dfrac{N_0 f^2}{A_c^2}, & |f| \leq W \\ 0, & \text{otherwise.} \end{cases} \tag{3.96}$$

Figure 3.32 summarizes the above derivations.

We are ready to obtain

the average power of the output noise $= \dfrac{N_0}{A_c^2} \displaystyle\int_{-W}^{W} f^2\, df = \dfrac{2N_0 W^3}{3 A_c^2}.$

$$\tag{3.97}$$

Note that the average output noise power is inversely proportional to the average carrier power. Consequently, in FM systems, increasing the carrier power has a *noise-quieting effect*.

Provided high CNR, the output SNR is given by

$$(SNR)_{\text{O,FM}} = \dfrac{3 A_c^2 k_f^2 P}{2 N_0 W^3}. \tag{3.98}$$

and the channel SNR is given by

$$(SNR)_{\text{C,FM}} = \dfrac{A_c^2}{2 W N_0}. \tag{3.99}$$

The figure of merit is thus given by

$$\left. \dfrac{(SNR)_O}{(SNR)_C} \right|_{\text{FM}} \simeq \dfrac{3 k_f^2 P}{W^2}. \tag{3.100}$$

When the CNR is high, increasing transmission bandwidth B_T implies better figure of merit for FM.

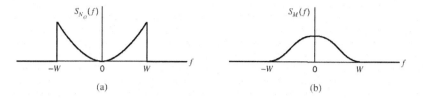

Fig. 3.33 Power spectral density of: (a) noise at the FM receiver output and (b) a typical message.

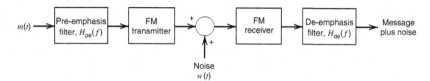

Fig. 3.34 Pre-emphasis and de-emphasis in FM.

We may observe from the above derivations that FM has the capability of minimizing effects from unwanted signals (or interferences). However, this holds only when the interference is weaker than the desired FM signal. Once the strength of the interference is close to that of the FM signal, the receiver could fluctuate back and forth, which is known as *capture effect*. We may experience capture effect when hearing music from FM broadcasting when driving a car. FM can also have threshold effect and can be improved by using a PLL.

Finally, recall from (3.96) that the noise has a psd as shown in Figure 3.33(a). However, typical message has psd as Figure 3.33(b). That is, noise power is high near the edges of transmission bandwidth but the signal is usually small in those edges (higher frequency range, which may have a lot of signal features such as audio and video). To resolve this dilemma, we can adopt the method as shown in Figure 3.34 showing, and the filters satisfy the following condition:

$$H_{de}(f) = \frac{1}{H_{pe}(f)}, \quad |f| \le W. \qquad (3.101)$$

3.9 Phase Locked Loops

As we have mentioned earlier direct FM demodulation is usually realized through a PLL. We are now to introduce more principles of

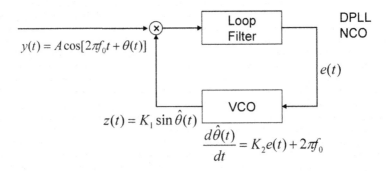

Fig. 3.35 Phase locked loop (PLL).

PLL, which are useful in many communications, computers, and IC design applications. PLL can be implemented as an analog loop or a digital loop. Here, we focus on the analog loop and leave the DPLL as an advanced subject. As shown in Figure 3.35, a PLL consists of a

- Mixer, to mix two waveforms (input and locally generated reference), and it can be generally considered as a PD;
- VCO, to modify waveform angle based on the input bias;
- Loop filter, to filter appropriate signal as the output and as the driving of VCO.

From Figure 3.35, the output of VCO

$$z(t) = K_1 \sin \hat{\theta}(t), \tag{3.102}$$

where the instantaneous frequency of $z(t)$ is given by

$$\frac{d\hat{\theta}(t)}{dt} = k_2 e(t) + 2\pi f_0. \tag{3.103}$$

The loop filter $f(t)$, an LPF, filters out the second harmonics of $y(t)$, $z(t)$. Consider a noise-free signal into the PLL

$$y(t) = A \cos[2\pi f_0 t + \theta(t)]. \tag{3.104}$$

Then

$$e(t) = \frac{AK_1 K_2}{2} \int_0^t f(t - \sigma) \sin[\theta(\sigma) - \hat{\theta}(\sigma)] d\sigma. \tag{3.105}$$

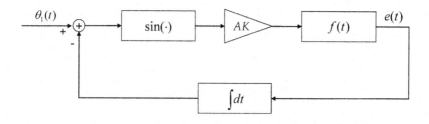

Fig. 3.36 Equivalent baseband PLL model.

The operation of PLL can be described by

$$\frac{d\hat{\theta}}{dt} = 2\pi f_0 t + \frac{K_2}{2} \int_0^t f(t-\sigma) A K_1 \sin[\theta(\sigma) - \hat{\theta}(\sigma)] d\sigma. \qquad (3.106)$$

Defining $\phi(t) \triangleq \theta(t) - \hat{\theta}(t)$ as the phase error, $\hat{\theta}_2(t) = \hat{\theta}(t) - 2\pi f_0 t$, and the loop gain $K = \frac{K_1 K_2}{2}$

$$\frac{d\phi(t)}{dt} = \frac{d\theta(t)}{dt} - 2\pi f_0 t - AK \int_0^t f(t-\sigma) \sin\phi(\sigma) d\sigma. \qquad (3.107)$$

Such a PLL is an excellent approximation to $\theta(t)$. The equivalent baseband formulation of the PLL is $(\theta_1(t) = \theta(t) - 2\pi f_0 t)$ Figure 3.36.

Now, we consider the operation of the first-order PLL, where $f(t) = \delta(t)$:

$$\theta(t) = 2\pi f_c t + \theta_c. \qquad (3.108)$$

From (3.107), we can derive the following equation:

$$\frac{d\phi(t)}{dt} = \frac{d\theta}{dt} - 2\pi f_0 t - AK \sin\phi(t)$$
$$= 2\pi(f_c - f_0) - AK \sin\phi(t). \qquad (3.109)$$

If $AK > 2\pi(f_c - f_0)$, the loop has stable points of operation at $\theta_0 \pm 2n\pi$, since $\phi = 0$ will have a stable solution at $\sin^{-1}[2\pi(f_c - f_0)/AK]$. Therefore, we have

$$2\pi(f_c - f_0) < AK. \qquad (3.110)$$

The above equation gives the pull-in/hold-in (or lock-in) range for a first-order PLL, i.e., the range of frequency deviation of the incoming

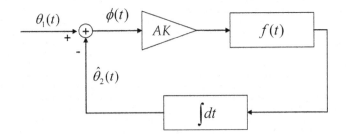

Fig. 3.37 Linear model of PLL.

carrier from the center frequency of the VCO for which it is possible to track. It also gives the lock-in range, i.e., the range of frequency variation for which the incoming angle/phase can be acquired.

We may have further understanding on PLL by considering the linearized loop model, which is valid for small loop error. In such a case, $\sin\phi(t) \approx \phi(t)$ Figure 3.37:

$$\frac{d\phi(t)}{dt} = \frac{d\theta_1(t)}{dt} - AK \int_0^t f(t-u)\phi(u)\,du. \qquad (3.111)$$

To further analyze, we proceed by taking the Laplace transform

$$s\Phi(s) = s\theta_1(s) - AKF(s)\Phi(s), \qquad (3.112)$$

where

$$\Phi(s) = L[\phi(t)]$$
$$\theta_1(s) = L[\theta_1(t)]$$
$$F(s) = L[f(t)].$$

It is straightforward to reach

$$\Phi(s) = \frac{\theta_1(s)}{1 + \frac{AKF(s)}{s}}. \qquad (3.113)$$

Or, we can equivalently reach the closed-loop transfer function

$$H(s) = \frac{\hat{\theta}_2(s)}{\hat{\theta}_2(s)} = \frac{\frac{AKF(s)}{s}}{1 + \frac{AKF(s)}{s}}. \qquad (3.114)$$

For the first-order loop, $F(s) = 1$ and input $\theta(f) = 2\pi f + \theta_0$ result in

$$\theta(s) = \frac{2\pi(f - f_0)}{s^2} + \frac{\theta_0}{s} \tag{3.115}$$

$$\Phi(s) = \frac{2\pi(f - f_0)}{s(s + AK)} + \frac{\theta_0}{s + AK}. \tag{3.116}$$

Using the steady-state property of Laplace transform, we have

$$\lim_{t \to \infty} \phi(t) = \lim_{s \to \infty} \Phi(s) = \frac{2\pi(f - f_0)}{AK}. \tag{3.117}$$

If $F(s) = 1 + \frac{a}{s}$ (i.e., second-order loop), we can obtain

$$\Phi(s) = \frac{2\pi(f - f_0) + \theta_0 s}{s^2 + AKs + aAK}. \tag{3.118}$$

Loop filter is an LPF that responds only to low-frequency component(s) and removes the double-frequency component(s). This filter is usually selected to have the relatively simple transfer function

$$G(s) = \frac{1 + \tau_2 s}{1 + \tau_1 s}, \tag{3.119}$$

where τ_1 and τ_2 are design parameters ($\tau_1 \gg \tau_2$) that control the bandwidth of the loop. A higher-order loop filter that contains additional poles may be used if necessary to obtain a better loop response. (For example, to track the signal under Doppler effect.)

The output of the loop filter provides the control voltage $v(t)$ for the VCO. VCO is basically a sinusoidal signal generator with an instantaneous phase

$$2\pi f_g t + \hat{\phi}(t) = 2\pi f_g t + K \int_{-\infty}^{t} v(\tau) \, d\tau, \tag{3.120}$$

where K is a constant gain.

In normal operation when the loop is tracking the phase of the incoming carrier, the phase error $\phi - \hat{\phi}$ is small, and thus

$$\sin(\hat{\phi} - \phi) \approx \hat{\phi} - \phi. \tag{3.121}$$

With this approximation, the PLL becomes linear and we can use a closed-loop system as shown in Figure 3.38 to represent the PLL. This

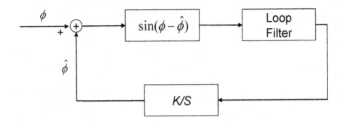

Fig. 3.38 Closed-loop model of PLL.

closed-loop model has a transfer function of

$$H(s) = \frac{\frac{KG(s)}{s}}{1 + \frac{KG(s)}{s}}. \tag{3.122}$$

If we adopt the simple loop filter of (3.119), then the closed-loop transfer function has the form

$$H(s) = \frac{1 + \tau_2 s}{1 + \left(\tau_2 + \frac{1}{K}\right)s + \frac{\tau_1}{K}s^2}. \tag{3.123}$$

Hence, the closed-loop system for the linearized PLL is of second-order. τ_2 controls the position of the zero. τ_1 and K are used to control the position of the closed-loop system poles. They are critical parameters when we design circuits to realize analog (and also digital) communication systems.

It is customary to express the denominator of $H(s)$ in a canonical form for the study of PLL

$$D(s) = s^2 + 2\xi w_n s + w_n^2, \tag{3.124}$$

where ξ is the loop damping factor and w_n is the natural frequency of the loop. Then, the closed-loop transfer function of a linear PLL model is given by

$$H(s) = \frac{\left(2\xi w_n - \frac{w_n^2}{K}\right)s + w_n^2}{s^2 + 2\xi w_n s + w_n^2}. \tag{3.125}$$

Based on (3.125), one-side noise equivalent bandwidth of the loop is therefore given by

$$B_{eq} = \frac{1 + (\tau_2 w_n)^2}{8\xi w_n} \tag{3.126}$$

We usually design a PLL with a small equivalent bandwidth in (3.126) so that we can increase the equivalent *SNR* in the PLL for effective operation, provided under the pull-in range. More detailed noise analysis of PLL is complicated and a good suggested reading is the book authored by A. Viterbi, "Principles of Coherent Communications," 1968.

Exercises

1. Show that SSB may be considered as a special case of VSB modulation by taking $f_v = 0$.
2. For an SSB receiver, find the
   (a) output *SNR*,
   (b) channel *SNR*,
   (c) figure of merit for the SSB receiver, which is the same as DSB-SC.

3. Let $\Phi(t)$ denote a zero-mean stationary Gaussian process with autocorrelation function $R_\Phi(\tau)$. This process modulates a sinusoidal to generate an angle-modulated process $\Omega(t) = A_c \cos[2\pi f_c t + \Phi(t)]$.
   (a) Show that $\Omega(t)$ is not stationary.
   (b) Show that the average autocorrelation function of $\Omega(t)$ is

   $$\bar{R}_\Omega(\tau) = \lim_{T \to \infty} \int_{-T/2}^{T/2} R_\Omega(t, t + \tau) \, dt$$
   $$= \frac{A_c^2}{2} \cos(2\pi f_c t) e^{-[R_\Phi(0) - R_\Phi(\tau)]}.$$

   (c) Find the psd of $\Omega(t)$ and find the bandwidth of $\Omega(t)$ (in terms of the bandwidth of $e^{R_\Phi(t)}$).

4. Using message signal $m(t) = 1/(1 + t^2)$, determine and sketch the modulated waveform for the following modulation methods:
   (a) AM with 50% modulation;
   (b) DSB-SC modulation;

(c) SSB modulation with only upper sideband transmitted and SSB modulation with only lower sideband transmitted;

(d) If we want to encode such a waveform by delta modulation, what is the major problem? How can we improve this problem?

5. The following figure shows the circuit diagram of a *square-law modulator*. The signal applied to the nonlinear device is relatively weak, such that it can be represented by a square law:

$$v_2(t) = a_1 v_1(t) + a_2 v_1^2(t),$$

where a_1 and a_2 are constants, $v_1(t)$ is the input voltage, and $v_2(t)$ is the output voltage. The input voltage is defined as

$$v_1(t) = A_c \cos(2\pi f_c t) + m(t),$$

where $m(t)$ is a message signal and $A_c \cos(2\pi f_c t)$ is the carrier wave.

(a) Evaluate the output voltage $v_2(t)$.

(b) Specify the frequency response that the tuned circuit in the figure must satisfy in order to generate an AM signal with f_c as the carrier frequency.

(c) What is the amplitude sensitivity of the AM signal?

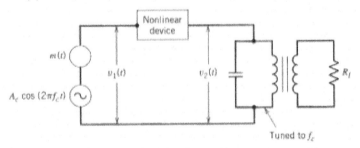

6. The following figure shows the circuit diagram of a *switching modulator*. Assume that the carrier wave $c(t)$ applied to the diode is large in amplitude, so that the diode acts like an ideal switch: it represents zero impedance when forward biased

(i.e., $c(t) > 0$). We may thus approximate the transfer characteristic of the diode–load resistor combination by a piecewise-linear characteristic defined as

$$v_2(t) = \begin{cases} v_1(t), & c(t) > 0 \\ 0, & c(t) < 0. \end{cases}$$

That is, the load voltage $v_2(t)$ varies periodically between the values $v_1(t)$ and zeros at a rate equal to the carrier frequency f_c. Hence, we may write

$$v_2(t) \cong [A_c \cos(2\pi f_c t) + m(t)]g_{T_0}(t),$$

where $g_{T_0}(t)$ is a periodic pulse train defined as

$$g_{T_0}(t) = \frac{1}{2} + \frac{2}{\pi}\sum_{n=1}^{\infty} \frac{(-1)^{n-1}}{2n-1}\cos[2\pi f_c t(2n-1)].$$

   (a) Find the AM wave component contained in the output voltage $v_2(t)$.

   (b) Specify the unwanted components in $v_2(t)$ that need to be removed by a BPF of suitable design.

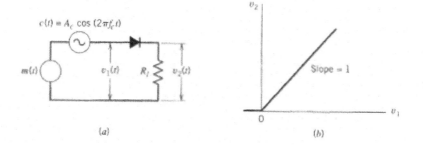

$c(t) = A_c \cos (2\pi f_c t)$

$m(t)$ $v_1(t)$ R_l $v_2(t)$ Slope = 1

(a)   (b)

7. The following figure shows the circuit diagram of a *balanced modulator*. The input applied to the top AM modulator is $m(t)$, whereas that applied to the lower AM modulator is $-m(t)$; these two modulators have the same amplitude sensitivity. Show that the output $s(t)$ of the balanced modulator consists of a DSB-SC modulated signal.

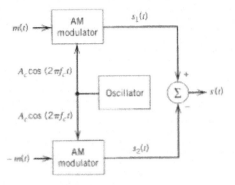

8. The following figure shows the block diagram of Weaver's method for generating SSB modulated waves. The message (modulating) signal $m(t)$ is limited to the band $f_a \leq |f| \leq f_b$. The auxiliary carrier applied to the first pair of product modulators has a frequency f_0, which lies at the center of this band, as shown by

$$f_0 = \frac{f_a + f_b}{2}.$$

The LPFs in the in-phase and quadrature channels are identical, each with a cutoff frequency equal to $(f_b - f_a)/2$. The carrier applied to the second pair of product modulators has a frequency f_c that is greater than $(f_b - f_a)/2$. Sketch the spectra at the various points in the modulator of below given figure and hence show that

   (a) For the lower sideband, the contributions of the in-phase and quadrature channels are of opposite polarity, and by adding them at the modulator output, the lower sideband is suppressed.

   (b) For the upper sideband, the contributions of the in-phase and quadrature channels are of the same polarity, and by adding them, the upper sideband is transmitted.

   (c) How would you modify the modulator in the figure so that the lower sideband is transmitted?

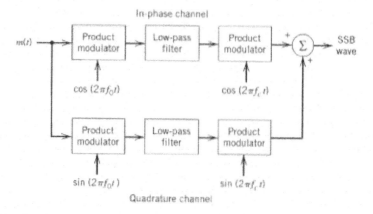

In-phase channel

Quadrature channel

9. The following block diagram of a real-time *spectrum analyzer* works on the principle of FM. The given signal $g(t)$ and a frequency-modulated signal $s(t)$ are applied to the multiplier and the output $g(t)s(t)$ is fed into a filter of impulse response $h(t)$. The $s(t)$ and $h(t)$ are *linear FM signals* whose instantaneous frequencies vary linearly with time at opposite rates, as shown by

$$s(t) = \cos(2\pi f_c t - \pi k t^2)$$
$$h(t) = \cos(2\pi f_c t + \pi k t^2),$$

where k is a constant. Show that the envelope of the filter output is proportional to the magnitude spectrum of the input signal $g(t)$ with kt playing the role of frequency f. Hint: use the complex notations for the analysis of band-pass signals and BPFs.

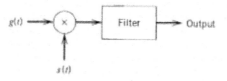

10. The FM signal

$$s(t) = A_c \cos\left[2\pi f_c t + 2\pi k_f \int_0^t m(\tau)d\tau\right]$$

is applied to the system shown in the below given figure consisting of a high-pass RC filter and an envelope detector. Assume that (a) the resistance R is small compared with the resistance of the capacitor C for all significant frequency components of $s(t)$, and (b) the envelope detector does not load the filter. Determine the resulting signal at the envelope detector output, assuming that $k_f |m(t)| < f_c$ for all t.

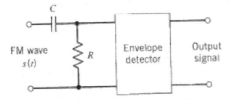

11. The *signal sideband version of angle modulation* is defined as

$$s(t) = \exp[-\hat{\phi}(t)]\cos[2\pi f_c t + \phi(t)],$$

where $\hat{\phi}(t)$ is the Hilbert transform of the phase function $\phi(t)$, and f_c is the carrier frequency.

(a) Show that the spectrum of the modulated signal $s(t)$ contains no frequency components in the interval $-f_c < f < f_c$ and is of infinite extent.

(b) Given that the phase function

$$\phi(t) = \beta \sin(2\pi f_m t),$$

where β is the modulation index and f_m is the modulation frequency, derive the corresponding expression for the modulated wave $s(t)$.

12. Consider the output of an envelope detector, which is reproduced here for convenience

$$y(t) = \{[A_c + A_c k_a m(t) + n_I(t)]^2 + n_Q^2(t)\}^{1/2}.$$

(a) Assume that the probability of the event

$$|n_Q(t)| > \varepsilon A_c |1 + k_a m(t)|$$

is equal to or less than δ_1, where $\varepsilon \ll 1$. What is the probability that the effect of the quadrature component $n_Q(t)$ is negligible?

(b) Suppose that k_a is adjusted related to the message signal $m(t)$ such that the probability of the event

$$A_c[1 + k_a m(t)] + n_I(t) < 0$$

is equal to δ_2. What is the probability that the estimation

$$y(t) \cong A_c[1 + k_a m(t)] + n_I(t)$$

is valid?

(c) Comment on the significance of the result in part (b) for the case when both δ_1 and δ_2 are small compared with unity.

13. Let R denote the random variable obtained by observing the output of an envelope detector at some fixed time. Intuitively, the envelope detector is expected to be operating well into the threshold region if the probability that the random variable R exceeds the carrier amplitude A_c is 0.5. On the other hand, if this same probability is 0.01, the envelope detector is expected to be relatively free of loss of message and the threshold effect.

(a) Assuming that the narrowband noise at the detector input is white, zero-mean, Gaussian with spectral density $N_0/2$ and the message bandwidth is W, show that the probability of the event $R \geq A_c$ is

$$P(R \geq A_c) = \exp(-\rho),$$

where ρ is the *CNR*, which is given by

$$\rho = \frac{A_c^2}{4W N_0}.$$

(b) Using the formula for this probability, calculate the *CNR* when (1) the envelope detector is expected to be well into the threshold region and (2) it is expected to be operating satisfactorily.

14. Suppose that the transfer functions of the pre-emphasis and de-emphasis filters of an FM system are scaled as follows:

$$H_{\text{pe}}(f) = k\left(1 + \frac{jf}{f_0}\right)$$

and

$$H_{\text{de}}(f) = \frac{1}{k}\left(\frac{1}{1 + jf/f_0}\right).$$

The scaling factor k is to be chosen so that the average power of the emphasized message signal is the same as that of the original message signal $m(t)$.

(a) Find the value of k that satisfies this requirement for the case when the psd of the message signal $m(t)$ is

$$S_M(f) = \begin{cases} \dfrac{S_0}{1 + (f/f_0)^2}, & -W \le f \le W \\ 0, & \text{elsewhere.} \end{cases}$$

(b) What is the corresponding value of the improvement factor I produced by using this pair of pre-emphasis and de-emphasis filters?

4

Pulse Modulation and Digital Coding

One major breakthrough in communications is to adopt digital signaling facilitating digital communications. Generally speaking, the advantages of digital communications over analog communications are as follows:

- Digital signaling/modulation methods have better resistance to noise.
- For long-haul communications using repeaters (such as intercontinental cable and microwave communications) to extend communication range, digital signaling can be regenerated for active repeaters, whereas analog signaling can be amplified only for passive repeaters. Digital signaling may have better overall performance in such application scenarios.
- Furthermore, digital signaling can apply error-correcting codes and error control techniques to resist channel noise and imperfections, so that information can be better protected.

A digital communication system usually consists of the following blocks as shown in Figure 4.1, while the upper part of the figure belongs to the transmitter and the lower part of the figure belongs to the receiver. The transmitter is composed of

- A signal (analog) waveform source and analog-to-digital conversion (ADC), whose purpose is to translate an analog waveform into digital (discrete time) samples;
- An encoder that includes source encoder and channel encoder, which are defined as information theory, and whose purpose is to translate digital samples into bit stream;

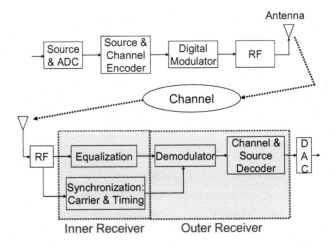

Fig. 4.1 Block diagram of a typical digital communication system.

- A digital modulator to modulate binary bit(s) into symbols for transmission;
- An RF section to translate a modulated waveform into appropriate frequency range, which is not the scope of this book.

Note that the channel coding and source coding are well-studied information theory and we only briefly introduce in this book.

The receiver, excluding the RF to down-convert the received waveform into appropriate IF or baseband range, can be generally divided into an inner receiver and an outer receiver for the digital realization portion. The outer receiver functions as the reverse operation of a transmitter to execute demodulation and then decoding for both channel coding and source coding. The major goal of the inner receiver is to extract useful signal parameters and to execute appropriate signal filtering, so that the outer receiver of demodulation and decoding can be achieved smoothly. The primary scope of this book is the outer receiver, especially the demodulation, whose principles can be very useful to other parts. Of course, we may observe the importance of receiver design in communications, regardless of analog or digital.

In what follows, we start digital communication system design from the description of technology converting analog information waveforms into discrete (or digital) pulse signaling at baseband, which can be used for baseband modulation in some cases too.

Generally speaking, such a process consists of the following two important steps:

(a) ADC to convert an analog waveform into a digital signaling format (i.e., samples with binary representation);

(b) Coding of a waveform to ensure minimum digits representing waveform and to better meet system performance.

4.1 Sampling

Sampling is usually considered as a discrete-time process to convert an analog waveform into a sequence of discrete values (known as *samples*). These samples are usually distributed in a uniform manner, but not necessarily uniform. To reach our purpose, we must be able to use such discrete-time samples to represent an analog waveform, and we must be able to reconstruct the analog waveform from these samples. The theoretical development is introduced as follows.

Let $g(t)$ an analog waveform with finite energy/power. The ideal sampled signal with the *sampling period* T_s is given by

$$g_\delta(t) = \sum_{n=-\infty}^{\infty} g(nT_s)\delta(t - nT_s), \tag{4.1}$$

where $\delta(t - nT_s)$ means a delta function at $t = nT_s$ and a sampling occurs at $t = nT_s$. If $\leftrightarrow$ denotes a Fourier transform pair, we can rewrite the above equation as

$$g_\delta(t) \leftrightarrow f_s \sum_{m=-\infty}^{\infty} G(f - mf_s), \tag{4.2}$$

where $g(t) \leftrightarrow G(f)$ and $f_s = 1/T_s$ is the sampling rate. Using the property of Fourier transform, we can get from (4.1)

$$G_\delta(f) = \sum_{n=-\infty}^{\infty} g(nT_s)\exp(-j2\pi nfT_s). \tag{4.3}$$

It is also known as *discrete-time Fourier transform*, which can be viewed as a complex Fourier series of the periodic function $G_\delta(f)$. Suppose that

$g(t)$ is strictly band-limited without any frequency component higher than W Hz. If we select sampling period $T_s = 1/2W$, then

$$G_\delta(f) = \sum_{n=-\infty}^{\infty} g\left(\frac{n}{2W}\right) \exp\left(-\frac{j\pi n f}{W}\right). \tag{4.4}$$

Using (4.2), we can express

$$G_\delta(f) = f_s G(f) + f_s \sum_{\substack{m=-\infty \\ m \neq 0}}^{\infty} G(f - m f_s). \tag{4.5}$$

Consequently, with the following two conditions:

(a) $G(f) = 0$ for $|f| \leq W$.
(b) $f_s = 2W$.

We can derive

$$G(f) = \frac{1}{2W} G_\delta(f), \quad |f| < W. \tag{4.6}$$

We can also re-write as the following equation:

$$G(f) = \frac{1}{2W} \sum_{n=-\infty}^{\infty} g\left(\frac{n}{2W}\right) \exp\left(-\frac{j\pi n f}{W}\right), \quad |f| < W. \tag{4.7}$$

We are ready to draw an important conclusion: if the sample values $g(n/2W)$ of a band-limited signal $g(t)$ for all n, then $G(f)$ is uniquely determined by the discrete-time Fourier transform (4.7). Furthermore, this sequence of $g(n/2W)$ contain all information in $g(t)$. Once we have the representation formula, we shall consider the signal reconstruction from samples. To obtain $g(t)$ from inverse Fourier transform, we have

$$g(t) = \int_{-\infty}^{\infty} G(f) \exp(j2\pi f t) df$$

$$= \int_{-W}^{W} \frac{1}{2W} \sum_{n=-\infty}^{\infty} g\left(\frac{n}{2W}\right) \exp\left(-\frac{j\pi n f}{W}\right) \exp(j2\pi f t) df. \tag{4.8}$$

By exchanging the operation of integration and summation under proper mathematical condition, we have

$$g(t) = \sum_{n=-\infty}^{\infty} g\left(\frac{n}{2W}\right) \frac{1}{2W} \int_{-W}^{W} \exp\left[j2\pi f\left(t - \frac{n}{2W}\right)\right] df. \tag{4.9}$$

Then, we are ready to reach the following theorem.

Theorem 4.1 (Interpolation Formula). To provide reconstruction of original signal $g(t)$ from the sequence of samples, we have

$$g(t) = \sum_{n=-\infty}^{\infty} g\left(\frac{n}{2W}\right) \frac{\sin(2\pi Wt - n\pi)}{(2\pi Wt - n\pi)}$$

$$= \sum_{n=-\infty}^{\infty} g\left(\frac{n}{2W}\right) \text{sinc}(2Wt - n), \quad -\infty < t < \infty. \quad (4.10)$$

Here, sinc function can be considered as the basis of an interpolation function. We can summarize *sampling theorem* in the following two ways:

(a) A band-limited signal of finite energy with all frequency components at $|f| \leq W$ can be completely described by the sampled values separated by $1/2W$.

(b) A band-limited signal of finite energy with all frequency components at $|f| \leq W$ can be completely recovered from its samples taken at the rate of $2W$ per second.

This sampling rate at $2W$ samples per second for signal bandwidth W is known as *Nyquist rate*. In fact, the process of uniform sampling a finite-energy continuous-time signal results in a periodic spectrum with a period equal to the sampling rate. In other words, the sampling process repeats the original spectrum in frequency as shown in Figure 4.2.

In case of under-sampling (sampling rate smaller than the Nyquist rate), the signal spectrum would be as shown in Figure 4.3, which is known as *aliasing* to result in signal un-recovered distortion. To avoid

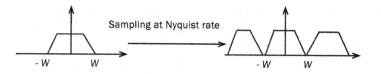

Fig. 4.2 Spectrum repetition of a band-limited sampled signal.

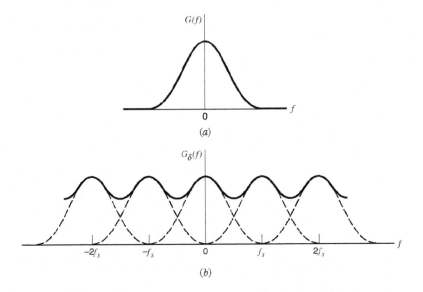

Fig. 4.3 (a) Signal spectrum; (b) aliasing.

aliasing,

(a) Low-pass filtering (anti-aliasing filter) is used to remove the high-frequency components (higher than Nyquist rate) that are not essential.

(b) The filtered signal is sampled at a rate slightly greater than the Nyquist rate.

4.2   Pulse-Amplitude Modulation (PAM)

The most straightforward way to give samples right representing values might be PAM, while the amplitudes of regularly spaced pulses are varied proportional to the corresponding sample values from a continuous message signal. Figure 4.4 depicts such a mechanism.

The generation of PAM consists of the following two steps:

- Instantaneous sampling of the message signal $m(t)$ with sampling period T_s satisfying sampling theorem.
- Pulse duration is kept constant $T < T_s$.

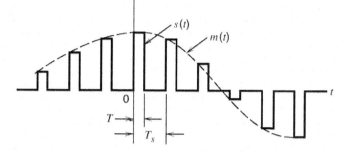

Fig. 4.4 PAM.

The above two steps can be realized by circuits known as *sample and hold*. Now, let $s(t)$ denote the generated pulse sequence as shown in Figure 4.4, which is expressed as

$$s(t) = \sum_{n=-\infty}^{\infty} m(nT_s) h(t - nT_s), \qquad (4.11)$$

where $m(nT_s)$ is the sample value of $m(t)$ at $t = nT_s$. $h(t)$ now represents a rectangular pulse with unitary amplitude and duration T, and is expressed as

$$h(t) = \begin{cases} 1, & 0 < t < T \\ \frac{1}{2}, & t = 0, \quad t = T \\ 0, & \text{otherwise.} \end{cases} \qquad (4.12)$$

The sampled version of $m(t)$ is given as

$$m_\delta(t) = \sum_{n=-\infty}^{\infty} m(nT_s)\delta(t - nT_s). \qquad (4.13)$$

Convolving $m_\delta(t)$ with pulse $h(t)$, we have

$$m_\delta(t) * h(t) = \int_{-\infty}^{\infty} m_\delta(\tau) h(t - \tau) \, d\tau$$

$$= \int_{-\infty}^{\infty} \sum_{n=-\infty}^{\infty} m(nT_s)\delta(\tau - nT_s) h(t - \tau) \, d\tau$$

$$= \sum_{n=-\infty}^{\infty} m(nT_s) \int_{-\infty}^{\infty} \delta(\tau - nT_s) h(t - \tau) \, d\tau. \qquad (4.14)$$

From the property of delta function, we obtain

$$m_\delta(t) * h(t) = \sum_{n=-\infty}^{\infty} m(nT_s)h(t - nT_s). \qquad (4.15)$$

From (4.11) and (4.15), in time domain and frequency domain,

$$s(t) = m_\delta(t) * h(t) \qquad (4.16a)$$

and

$$S(f) = M_\delta(f)H(f), \qquad (4.16b)$$

where $s(t) \leftrightarrow S(f)$, $m_\delta(t) \leftrightarrow M_\delta(f)$, $h(t) \leftrightarrow H(f)$. Since

$$M_\delta(f) = f_s \sum_{k=-\infty}^{\infty} M(f - kf_s), \qquad (4.17)$$

we therefore have

$$S(f) = f_s \sum_{k=-\infty}^{\infty} M(f - kf_s)H(f). \qquad (4.18)$$

Here, we have to consider how to reconstruct the message signal $m(t)$ from PAM signal $s(t)$. Note that the Fourier transform of rectangular pulse $h(t)$ is given by

$$H(f) = T \operatorname{sinc}(fT) \exp(-j\pi fT). \qquad (4.19)$$

We may observe amplitude distortion as well as a delay of $T/2$ from the above equation. Such a distortion is called *aperture effect*. It can be resolved by introducing the equalizer after PAM reconstruction filter. Equalizer is widely used in communication system to "equalize" unwanted filtering effect. The desired filter response to equalize aperture effect is as follows:

$$\frac{1}{|H(f)|} = \frac{1}{T \operatorname{sinc}(fT)} = \frac{\pi f}{\sin(\pi fT)}. \qquad (4.20)$$

It is obvious to note the weakness of PAM susceptible to noise. There are other popular forms of pulse modulation other than PAM; they are

- *Pulse duration modulation* (PDM): It is also known as pulse-width modulation. The duration of a pulse is varied with the corresponding sample of message signal. The message information is thus embedded into the duration of pulse to be more robust against noise Figure 4.5.
- *Pulse position modulation* (PPM): The relative position of the pulse to the unmodulated pulse is varied with the corresponding sample of message signal. PPM can be considered as a variation from PDM by transforming the duration of the pulse into the position of fixed width pulse.

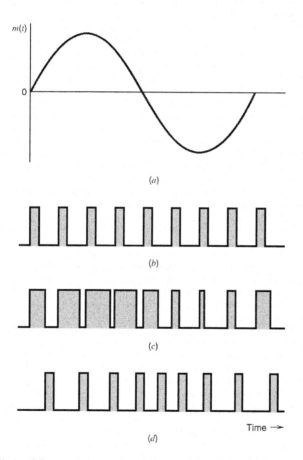

Fig. 4.5 (a) Sinusoidal message waveform; (b) equally spaced pulses; (c) PDM; and (d) PPM.

Between the above-mentioned pulse modulation, PPM has the most robustness against noise. It is also widely applied in optical communications, which are to be discussed later. Its bandwidth depends on the number of time slots (time resolution) for a symbol.

The widely applied concept in pulse modulation is to include coding structure, which is known as *pulse code modulation* (PCM). The generation of PCM involves two steps: *sampling* and *quantization*. Sampling gives the discrete-time representation of a message signal by a sequence of samples. Quantization gives digital representation of each discrete-time sample (usually on amplitude). PCM is the foundation of many today's voice, audio, image, video source coding schemes, which are used in so many state-of-the-art multimedia devices.

Figure 4.6 illustrates an example of PCM. The amplitude of each sample (at time $0, T_s, \ldots, nT_s, \ldots$) is encoded into 3 bits to represent amplitude information, which involves sapling and quantization steps. Within the figure, from time origin, we can obtain the PCM bit sequence as follows:

$$(000, 000, 001, 001, 001, 010, 011, 110, 111, 111, 101, 001, 000, 010, \ldots).$$

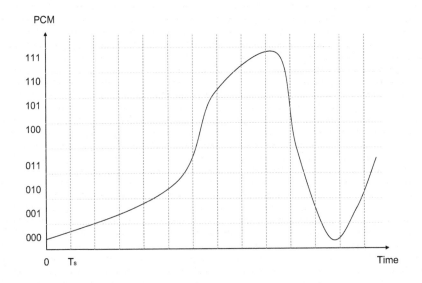

Fig. 4.6 PCM.

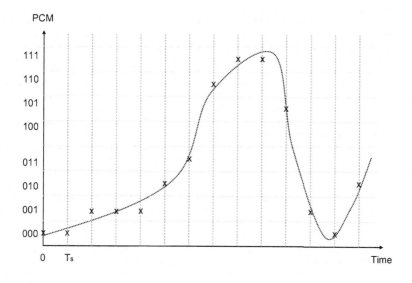

Fig. 4.7 PCM signal reconstruction.

Such a sequence can be used for reconstruction (or approximation) of the signal. Figure 4.7 illustrates the reconstruction of a signal based on the mid-point of each coded sample (marked as x in the figure). If we properly connect the curve based on those marks in the figure, we can observe almost the same waveform. The remaining error is called quantization error and will be discussed in the later section.

4.3   Quantization

Since the early days of information theory, there is a branch of information theory called source coding to explore the theoretical limit of converting an analog signal into effective bits. It is constructed based on the rate-distortion theory, and one important subject is quantization. Interested readers shall study more in relevant books.

At this point, we are interested in amplitude quantization, which is defined as the process to transform sampled amplitude $m(nT_s)$ of message signal $m(t)$ at $t = nT_s$, into a discrete amplitude $v(nT_s)$ taken from a finite set of amplitude *alphabets* (i.e., a finite set of possible amplitude values). Let us consider a memoryless quantizer as shown in Figure 4.8.

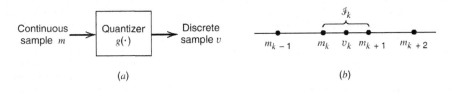

$$(a) \qquad\qquad\qquad\qquad (b)$$

Fig. 4.8 A memoryless quantizer.

The signal amplitude m is specified by an index k if the signal amplitude lies inside a partition interval, that is,

$$I_k : \{m_k < m \leq m_{k+1}\}, \quad k = 1, 2, \ldots, L, \qquad (4.21)$$

where L is the total number of possible amplitude levels adopted in the quantizer. The discrete amplitude values $m_k, k = 1, 2, \ldots, L$, at the quantizer input are known as *decision levels* or *decision thresholds*. At the quantizer output, the index k is mapped into amplitude v_k that represents all possible amplitudes within the interval $I_k \cdot v_k, k = 1, 2, \ldots, L$, are called representation levels or reconstruction levels, and the space between two adjacent levels is defined as the *step-size*. Consequently, the quantizer output $v = v_k$ if $m \in I_k$. The mapping (or function)

$$v = g(m) \qquad (4.22)$$

represents quantizer characteristics and is a stair-type function as shown in Figure 4.9 showing mid-tread and mid-rise types of quantizers.

A quantizer of uniformly spaced representation levels is called a *uniform quantizer*; otherwise, it is called a *non-uniform quantizer*. Of course, uniform quantization is practically preferred by considering implementation clock. Since there exists a difference between m and v, we call this difference as *quantization error* or *quantization noise*. For evaluation, it is reasonable to consider quantization input m from a zero-mean random variable M and to be mapped into a discrete random variable V by $g(\cdot)$.

Let the quantization error be

$$\begin{aligned} q &= m - v \\ Q &= M - V. \end{aligned} \qquad (4.23)$$

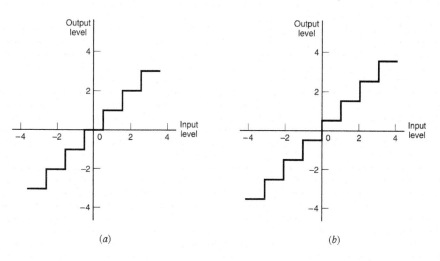

Fig. 4.9 Quantization: (a) Mid-tread; (b) mid-rise.

Suppose that the quantizer is symmetric as shown in Figure 4.9(b). With M having zero mean, V and Q are also having zero mean. Let the amplitude of m be in the range of $(-m_{max}, m_{max})$. Assuming a uniform quantizer of L representation levels, the step-size is given by

$$\Delta = \frac{2m_{max}}{L}. \tag{4.24}$$

For a uniform quantizer, the quantization error Q is distributed within $[-\Delta/2, \Delta/2]$. Furthermore, for small step-size, it is reasonable to assume Q to be uniformly distributed. That is, the probability density function of Q is given by

$$f_Q(q) = \begin{cases} \frac{1}{\Delta}, & -\frac{\Delta}{2} < q \le \frac{\Delta}{2} \\ 0, & \text{elsewhere} \end{cases} \tag{4.25}$$

and its variance is expressed as

$$\sigma_Q^2 = E[Q^2] = \int_{-\Delta/2}^{\Delta/2} q^2 f_Q(q) dq$$

$$= \frac{1}{\Delta} \int_{-\Delta/2}^{\Delta/2} q^2 \, dq = \frac{\Delta^2}{12}. \tag{4.26}$$

Denote the number of bits per sample to construct the binary codes as R and

$$L = 2^R$$
$$R = \log_2 L. \tag{4.27}$$

Substituting (4.27) into (4.24), we obtain the step-size

$$\Delta = \frac{2m_{\max}}{2^R}. \tag{4.28}$$

Therefore, the variance of quantization error (or, quantization noise power) is given by

$$\sigma_Q^2 = \frac{1}{3}m_{\max}^2 2^{-2R}. \tag{4.29}$$

Supposing P is the average power of the message signal, the output signal-to-noise ratio (SNR) of a uniform quantizer becomes

$$SNR_O = \frac{P}{\sigma_Q^2}$$

$$= \frac{3P}{m_{\max}^2}2^{2R}. \tag{4.30}$$

The output SNR of a uniform quantizer increases exponentially with increasing number of bits per sample, R, which is proportional to channel bandwidth B_T. When FM and AM are limited by receiver noise, such binary coded modulation systems are limited by the quantization noise.

We can further develop general optimization for scalar quantizers by defining distortion measure $d(m, v_k)$. The optimal condition for scalar quantizer is to minimize average distortion based on distortion measure, that is,

$$D = \sum_{k=1}^{L} \int_{m \in T_k} d(m, v_k) f_M(m) dm. \tag{4.31}$$

The most common distortion measure is mean-square distortion, which is given by

$$d(m, v_k) = (m - v_k)^2. \tag{4.32}$$

We may have the following two design approaches based on this measure:

(a) Optimal encoder for a given decoder: The quantizer is in fact to give the nearest neighbor as the solution.

(b) Optimal decoder for a given encoder: We can use the necessary condition to get optimality, which suggests the well-known Lloyd–Max quantizer, optimal achieved by iterations.

Now, let us get back to the most widely applied PCM after introducing quantization. Figure 4.10 depicts a typical PCM-based communication system. At the transmitter end, quantization and encoder are usually implemented in the same circuit, which is called *analog-to-digital converter* (ADC). In today's integrated circuits implementation, ADC is typically with digital baseband communication circuits, but ADC is a sort of mixed-mode circuits.

Practical applications of PCM are usually not using uniform quantization. For example, in voice or speech communications, loud signal can be 30 dB stronger than weaker signal. Consequently, non-uniform quantization is commonly employed. In telephone communication, it

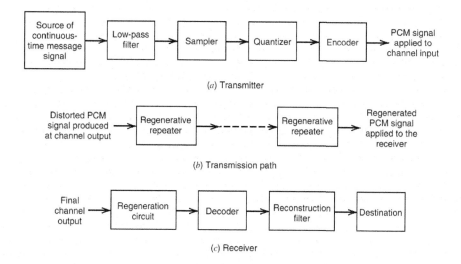

Fig. 4.10 Block diagram of a PCM system.

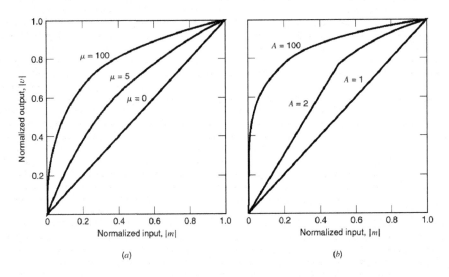

Fig. 4.11 Compression laws: (a) μ-law; (b) A-law.

is common to pass a baseband signal through a *compressor* prior to a uniform quantizer. For North American standards, μ-*law* is used in practice as the compression law and is defined as follows. Its IO relationship is also plotted in Figure 4.11(a) and is given by

$$|v| = \frac{\log(1 + \mu|m|)}{\log(1 + \mu)}. \tag{4.33}$$

We may observe that μ-law is approximately linear at low input levels ($\mu|m| \ll 1$) and is approximately logarithmic at high input levels ($\mu|m| \gg 1$). To restore the signal samples at the receiver, we need an inverse operating device called *expander*. This technique to combine a *compressor* and an *expander* is named as *compander*.

Another widely used (typical European systems) compression law is *A-law*, which is given as

$$|v| = \begin{cases} \dfrac{A|m|}{1 + \log A}, & 0 \le |m| \le \dfrac{1}{A} \\[3mm] \dfrac{1 + \log(A|m|)}{1 + \log A}, & \dfrac{1}{A} \le |m| \le 1. \end{cases} \tag{4.34}$$

Both μ-*law* and A-*law* are depicted in Figure 4.11.

The overall performance of a PCM system is clearly influenced by channel noise and quantization noise. Their impacts are to create bit errors at the received signal. It gives us a very important concept that the performance of digital communication systems (surely including PCM systems) is evaluated by *average probability of bit (or symbol) errors*, rather than just *SNR* in analog communications. Sometimes, we may also use *bit (or symbol) error rate* (BER) as the performance index, which is almost the same as the average probability of bit errors. As usual, we consider additive white Gaussian noise in channels, and thus we define E_b/N_0, which is the ratio of the transmitted signal energy per bit (E_b) to the noise power spectral density (N_0), within signal bandwidth. Note that one symbol may represent a few bits in communication signaling.

4.4 Time-Division Multiplexing (TDM)

The purpose of communications is not only for successful transmission of bits, but for successful reception of messages consisting of *frames* or *packets*. We start from some common format of line codes that are useful to define the pulse shape for the purpose of communications of bits. Figure 4.12 shows a few common line codes (or bit signaling formats) for binary data bits, including

- Unipolar non-return-to-zero (NRZ): It is in fact on–off signaling and is most widely used in digital communication signaling.
- Polar NRZ: It is the actual situation for coherent digital communication signaling.
- Unipolar return-to-zero (RZ): The attractive part is the spikes at $f = 0, \pm 1/T_b$, which is useful for synchronization.
- Bipolar RZ (BRZ): Equal amplitude positive pulses and negative pulses are used alternatively for symbol "1" and no pulse for "0." There is no DC term for BRZ. It is also known as *alternative mark inversion* signaling.
- Split-phase (Manchester code): DC term is suppressed and is also good for synchronization due to the existence of

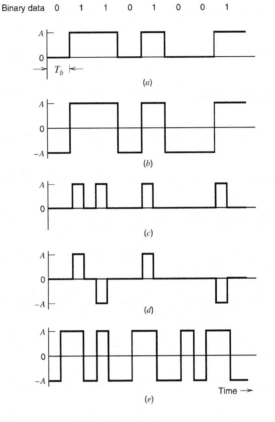

Fig. 4.12 Line codes for binary data.

zero-crossing in each symbol/bit, but at the price of double bandwidth.

Another method, *differential encoding*, to encode information is through signal transitions and is useful in many bandwidth efficient applications. Figure 4.13 demonstrates an example.

Then, we wish to bring a join utilization of a common communication channel by integrating independent message sources without mutual interference among them. For digital signaling, this purpose can be realized based on the time-share concept as shown in Figure 4.14,

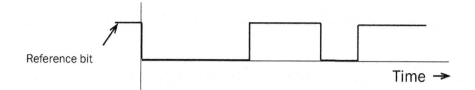

Reference bit

Time →

Fig. 4.13 Differential encoding.

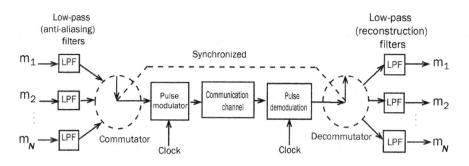

Fig. 4.14 Block diagram of a TDM system.

provided *synchronization* (timing alignment) among messages. It is called TDM, as a counterpart of frequency-division multiplexing.

Synchronization to align timing is obviously important in digital communications. Synchronization can be realized at different levels, such as bit level, symbol level, frame/packet level, etc. This can be further generalized to frequency domain for carrier and phase.

TDM makes *digital hierarchy* (DS) possible. It starts from DS0, globally common 64 kbps for PCM voice signals. It continues building up:

- DS1: TDM of 24 DS0 signals forms DS1 at 1.544 Mbps. T1 at the same rate is the typical system to carry DS1 signals.
- DS2: Four DS1 plus overhead results in DS2 at 6.312 Mbps.
- DS3: It combines 7 DS2 to form DS3 at 44.736 Mbps.
- DS4: Six DS3 signals result in DS4 at 274.176 Mbps.
- DS5: Combining 2 DS4 obtains DS5 at 560.16 Mbps.

4.5   Delta Modulation

Up to now, we may summarize the advantages of PCM as follows:

- Robustness to channel noise, interference, and impairments;
- Efficient *regeneration* of the coded signal along the transmission path to make long-distance communication very realizable;
- A uniform signal format for different kinds of signals;
- Easy to work with TDM and DS;
- Secure by incorporating encryption.

However, we still need *data compression* to save the communication bandwidth in many bandwidth-efficient applications. The challenge would be "do we need several bits to represent each sample?" The minimal number of bits for each sample would be 1, which shall be enough to build up a staircase approximation by using one bit per sample. This concept as shown in Figure 4.15 is called *delta modulation*. However, to ensure good performance, we may over-sample the message signal.

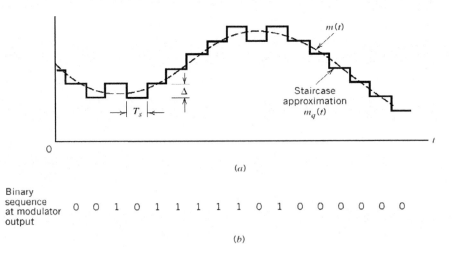

(*a*)

Binary
sequence
at modulator   0   0   1   0   1   1   1   1   1   0   1   0   0   0   0   0   0   0
output

(*b*)

Fig. 4.15 Concept of delta modulation.

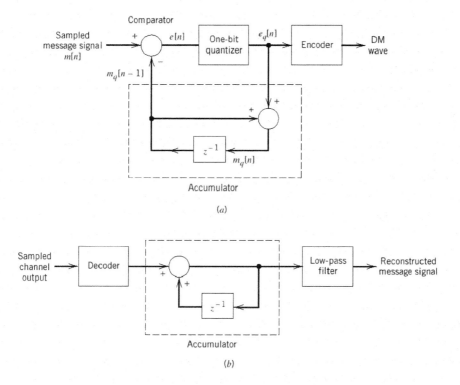

Fig. 4.16 Delta modulation system: (a) transmitter/encoder; (b) receiver/decoder.

The operation of delta modulation is depicted in Figure 4.16. Let $m_q(t)$ denote the staircase approximation of input message $m(t)$:

$$e[n] = m[n] - m_q[n-1], \tag{4.35}$$

$$e_q = \Delta \mathrm{sgn}(e[n]), \tag{4.36}$$

$$m_q[n] = m_q[n-1] + e_q[n], \tag{4.37}$$

where $e[n]$ is an error signal representing the difference between the current sample $m[n]$ and latest approximation $m_q[n-1]$; $e_q[n]$ is the quantized version of $e[n]$; the quantizer output $m_q[n]$ is encoded to generate DM signal. The one-bit quantizer is actually a *hard limiter*. The quantizer output is applied to the accumulator to produce

$$m_q[n] = \Delta \sum_{i=1}^{n} \mathrm{sgn}(e[i]) = \sum_{i=1}^{n} e_q[i]. \tag{4.38}$$

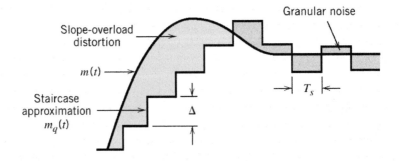

Fig. 4.17 Quantization error problems in delta modulation.

Delta modulation suffers from the following phenomenon shown in Figure 4.17:

- Slope-overload distortion: The signal changes so fast such that the delta approximation cannot catch up.
- Granular noise: The signal changes so slowly such that the step size becomes relatively large.

Both quantization error problems can be resolved by adaptive methodology that we will introduce later.

Finally, a variation of delta modulation called *sigma–delta modulation* (or delta–sigma modulation), Σ–Δ modulation, has been widely applied in modern ADC IC design. It over-samples and has an extra integration in front of delta modulation and another low-pass filter (or comb filter) to estimate message signal. Figure 4.18 depicts two equivalent models of Σ–Δ modulation.

4.6   Linear Prediction

Before introducing more mature source modulation schemes, we have to visit a concept, linear prediction, in more detail. Let us consider a *finite duration impulse response* (FIR) discrete-time filter as shown in Figure 4.19 with three functional blocks:

(a) p unit-delay elements;
(b) multipliers with weighting coefficients $w_1, w_2, \ldots, w_p$;

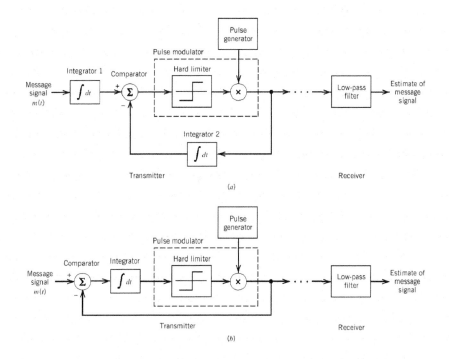

Fig. 4.18 Σ–Δ modulation.

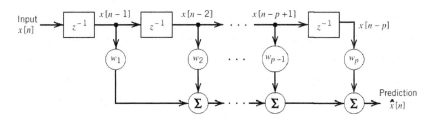

Fig. 4.19 Linear prediction filter of order p.

(c) adders to sum over the delayed inputs to produce the output $\hat{x}[n]$.

More precisely, $\hat{x}[n]$, the *linear prediction* of the input, is defined as the convolution sum

$$\hat{x}[n] = \sum_{k=1}^{p} w_k x[n-k]. \tag{4.39}$$

The *prediction error* is defined as

$$e[n] = x[n] - \hat{x}[n]. \tag{4.40}$$

Our goal is to select $w_1, w_2, \ldots, w_p$ in order to minimize the performance index J, and we use *mean-square error* here, which is given by

$$J = E[e^2[n]]. \tag{4.41}$$

The performance index is therefore

$$J = E[x^2[n]] - 2\sum_{k=1}^{p} w_k E[x[n]x[n-k]]$$

$$+ \sum_{j=1}^{p}\sum_{k=1}^{p} w_j w_k E[x[n-j]x[n-k]]. \tag{4.42}$$

The input signal $x(t)$ is assumed from the sample function of a stationary process $X(t)$ of zero mean, and thus $E[x[n]] = 0$ $\forall n$. We define

$$\sigma_X^2 = \text{variance of a sample of the process } X(t) \text{ at time } nT_s$$
$$= E[x^2[n]] - (E[x[n]])^2$$
$$= E[x^2[n]];$$

$$R_X(kT_s) = \text{autocorrelation of the process } X(t) \text{ for a lag of } kT_s$$
$$= R_X[k]$$
$$= E[x[n]x[n-k]].$$

(4.42) can be simplified as

$$J = \sigma_X^2 - 2\sum_{k=1}^{p} w_k R_X[k] + \sum_{j=1}^{p}\sum_{k=1}^{p} w_j w_k R_X[k-j]. \tag{4.43}$$

We can reach the necessary condition of optimality by differentiating filter coefficients, then we can obtain the famous *Wiener–Hopf* equation for linear prediction, that is,

$$\sum_{j=1}^{p} w_j R_X[k-j] = R_X[k] = R_X[-k], \quad k = 1, 2, \ldots, p. \tag{4.44}$$

Theorem 4.2 (Wiener–Hopf Equation in Matrix Form). Let

$$\mathbf{w_o} = p\text{-by-1 optimum coefficient vector}$$
$$= [w_1, w_2, \ldots, w_p]^{\mathrm{T}};$$

$$\mathbf{r_X} = p\text{-by-1 autocorrelation vector}$$
$$= [R_X[1], R_X[2], \ldots, R_X[p]]^{\mathrm{T}};$$

$$\mathbf{R_X} = p\text{-by-}p \text{ autocorrelation matrix}$$

$$= \begin{bmatrix} R_X[0] & R_X[1] & \cdots & R_X[p-1] \\ R_X[1] & R_X[0] & \cdots & R_X[p-2] \\ \vdots & \vdots & & \vdots \\ R_X[p-1] & R_X[p-2] & \cdots & R_X[0] \end{bmatrix}.$$

Then,

$$\mathbf{R_X w}_o = \mathbf{r_X}. \tag{4.45}$$

The optimal solution for coefficients is obtained from

$$\mathbf{w}_o = \mathbf{R_X}^{-1} \mathbf{r_X}. \tag{4.46}$$

The minimum mean-square error is given by

$$J_{\min} = \sigma_X^2 - \mathbf{r_X^T R_X^{-1} r_X}. \tag{4.47}$$

Note that $J_{\min} \leq \sigma_X^2$.

In real world, especially to encode signal waveforms, $R_X[k]$ in (4.46) is always varying, and we may use *adaptive prediction*, which recursively computes and adjust tap weights. Our goal is to find the minimum point of the bowl-shaped error surface, and the predictor is along the direction of steepest descent of the error surface. That is, we compute

$$g_k = \frac{\partial J}{\partial w_k}, \quad k = 1, 2, \ldots, p. \tag{4.48}$$

This is exactly the idea of the *method of steep descent*. Let $w_k[n]$ denote the value of the kth tap-weight at the nth iteration. Update is achieved by

$$w_k[n+1] = w_k[n] - \frac{1}{2}\mu g_k, \quad k = 1, 2, \ldots, p, \tag{4.49}$$

where μ is the step-size to control the speed of adaptation. By differentiating J with respect to w_k, we have

$$g_k = -2R_X[k] + 2\sum_{j=1}^{p} w_j R_X[k-j]$$

$$= -2E[x[n]x[n-k]] + 2\sum_{j=1}^{p} w_j E[x[n-j]x[n-k]], \quad k=1,2,\ldots,p$$

$$(4.50)$$

The estimate of g_k at the nth iteration is given by

$$\hat{g}_k[n] = -2x[n]x[n-k] + 2\sum_{j=1}^{p} w_j[n]x[n-j]x[n-k], \quad k=1,2,\ldots,p$$

$$(4.51)$$

From (4.51) and (4.49), by factoring common term $x[n-k]$, we may have

$$\hat{w}_k[n+1] = \hat{w}_k[n] + \mu x[n-k]\left(x[n] - \sum_{j=1}^{p} \hat{w}_j[n]x[n-j]\right)$$

$$= \hat{w}_k[n] + \mu x[n-k]e[n], \quad k=1,2,\ldots,p, \qquad (4.52)$$

where $e[n]$ is the prediction error and is expressed as

$$e[n] = x[n] - \sum_{j=1}^{p} \hat{w}_j[n]x[n-j]. \qquad (4.53)$$

Adaptive prediction can be realized as shown in Figure 4.20.

In order to enhance the communication efficiency by using less bits than PCM to transmit the same amount of information (that is important in bandwidth efficient communications), we may use *differential PCM* (DPCM) to encode the difference signal. DPCM operation is summarized in Figure 4.21. We usually design DPCM causing less slope overload and granular noise by encoding difference for more than 1 bit.

To result in even lower data rate but to keep slope overload and granular noise within reasonable level, *adaptive DPCM* (ADPCM) can be applied by dynamically adjusting the coefficients of the prediction filter. ADPCM and its variations are used in modern digital wireless communication while radio spectrum is cherished.

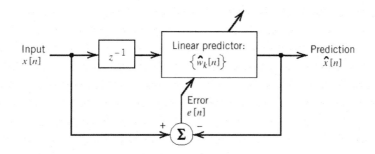

Fig. 4.20 Linear adaptive prediction.

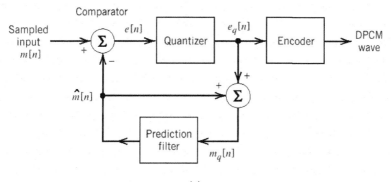

(a)

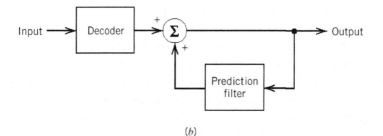

(b)

Fig. 4.21 Block diagram of DPCM.

4.7   Digital Coding of Waveforms

Based on the above technology, digital coding of waveforms is the fundamental technology in modern multimedia applications. It is not possible for us to introduce every application, and we use common examples to illustrate the usage. We start from the compact disk (CD) digital

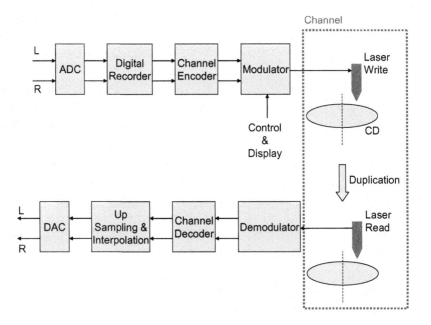

Fig. 4.22 CD digital audio.

audio system as shown in Figure 4.22, which form the foundation of many multimedia and storage systems. Laser write and laser read could be considered as a communication transmission and reception and thus consist of an equivalent communication channel. Such a communication also involves electro-optic conversion. The channel coding mechanism is pretty mature using a concatenated code consisting of convolutional code and Reed–Solomon code.

Another common digital coding of waveform is the image coding, while the most popular format is JPEG. Figure 4.23 shows the block diagram of JPEG encoder. An image is divided into nonoverlapping blocks, and each block has 8 × 8 pixels. Each block is transformed into frequency domain by a 2-D discrete cosine transform (DCT). The entropy coder consists of two stages. The first stage is either a predictive coder for DC or a run-length coder for AC. The second stage is usually a Huffman coder.

The video coding is based on the image coding, as we can treat video as a series of images at a very high rate that our eyes cannot tell.

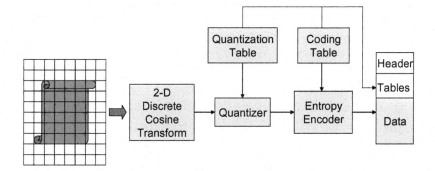

Fig. 4.23 JEPG encoder.

The most popular video coding is MPEG. The basic structure of MPEG consists of eight frames in a cycle, with the following three types:

- I-frame: They do not reference other pictures/images in the coded bit stream. The 1st and 8th frames are I-frames in an 8-frame cycle.
- P-frame: Predicted pictures are coded using motion-compensated prediction from past I-frames and P-frames. The 4th and 7th frames are P-frames.
- B-frame: Bidirectional predicted pictures using motion-compensated prediction from I-frames and P-frames have the highest degree of compression. The 2nd, 3rd, 5th, and 6th frames are B-frames.

There are quite a lot of further updated versions of MPEG, and MPEG-2 is widely used in digital TV. For better spectrum efficiency, more video coding schemes have been developed, such as H.264 and many others. Interested readers may find video coding references for further reading.

Exercises

1. In this problem we explore another method for the approximate realization of a matched filter, this time using the simple resistance–capacitance (RC) low-pass filter shown in the

following figure. The frequency response of this filter is

$$H(f) = \frac{1}{1 + jf/f_0},$$

where $f_0 = 1/2\pi RC$. The input signal $g(t)$ is a rectangular pulse of amplitude A and duration T. The requirement is to optimize the selection of the 3 dB cutoff frequency f_0 of the filter so that the peak pulse-to-signal ratio at the filter output is maximized. With this objective in mind, show that the optimum value f_0 is 0.2/T, for which the loss in *SNR* compared with the matched filter is about 1 dB.

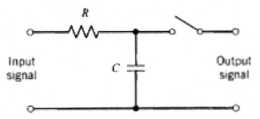

2. In a binary PCM system, symbols 0 and 1 have *a priori prob-abilities* p_0 and p_1, respectively. The conditional probability density function of the random variable Y (with sample value y) obtained by sampling the matched filter output in the receiver of the above figure at the end of a signaling interval, given that symbol 0 was transmitted, is denoted by $f_Y(y|0)$. Similarly, $f_Y(y|1)$ denotes the conditional probability density function of Y, given that symbol 1 was transmitted. Let λ denote the threshold used in the receiver, so that if the sample value y exceeds λ, the receiver decides in favor of symbol 1; otherwise, it decides in favor of symbol 0. Show that the optimum threshold λ_{opt}, for which the average probability of error is a minimum, is given by the solution of

$$\frac{f_Y(\lambda_{\text{opt}}|1)}{f_Y(\lambda_{\text{opt}}|0)} = \frac{p_0}{p_1}.$$

3. A binary PCM system using polar NRZ signaling operates just above the error threshold with an average probability of error equal to 10^{-6}. Suppose that the signaling rate is doubled. Find the new value of the average probability of error.

4. An analog signal is sampled, quantized, and encode into a binary PCM wave. The specifications of the PCM system include the following:

Sampling rate $= 8\,\text{kHz}$;

Number of representation levels $= 64$;

The PCM wave is transmitted over a baseband channel using discrete PAM. Determine the minimum bandwidth required for transmitting the PCM wave if each pulse is allowed to take on the following number of amplitude levels: 2, 4, or 8.

5. Two random variables X and Y are uniformly distributed on the square as shown in the following figure:

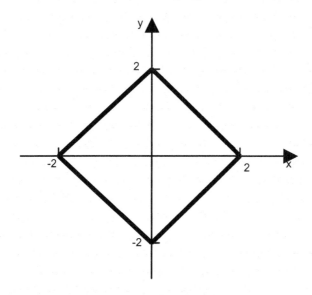

(a) Find $f_x(x)$ and $f_y(y)$.

(b) Assume that each of the random variables X and Y is quantized using four-level uniform quantizers. What is the resulting distortion? What is the resulting number of bits/pair (X, Y)?

(c) Now assume that instead of scalar quantizers of X and Y, we employ a vector quantizer with the same number of bits/source output pair (X, Y) as (b).

What is the resulting distortion for this vector quantizer?

6. The scheme shown in the following figure may be viewed as a differential encoder (consisting of the modulo-2 adder and the 1-unit delay element) connected in cascade with a special form of correlative coder (consisting of the 1-unit delay element and summer). A single-delay element is shown in the following figure since it is common to both the differential encoder and the correlative coder. In this differential encoder, a transition is represented by symbol 0 and no transition by symbol 1.

(a) Find the frequency response and impulse response of the correlative coder part of the scheme shown in the following figure.

(b) Show that this scheme may be used to convert the ON–OFF representation of a binary sequence (applied to the input) into the bipolar representation of the sequence at the output. You may illustrate this conversion by considering the sequence 010001101.

For description of ON–OFF, bipolar, and differential encoding of binary sequences, see Section 3.7.

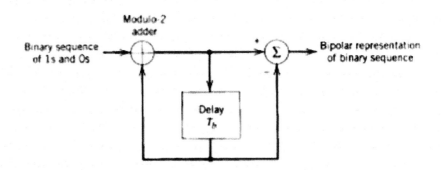

7. Consider a baseband M-ary system using M discrete amplitude levels. The receiver model is shown in the following figure, the

operation of which is governed by the following assumptions:

(a) The signal component in the received wave is given by

$$m(t) = \sum_n a_n \operatorname{sinc}\left(\frac{t}{T} - n\right),$$

where $1/T$ is the signaling rate in bauds.

(b) The amplitude levels are $a_n = \pm A/2, \pm 3A/2, \ldots,$ $\pm(M-1)A/2$ if M is even
, and $a_n = 0, \pm A, \ldots, \pm(M-1)A/2$ if M is odd.

(c) The M levels are equi-probable, and the symbols transmitted in adjacent time slots are statistically independent.

(d) The channel noise $w(t)$ is white and Gaussian with zero mean and power spectral density $N_0/2$.

(e) The low-pass filter is ideal with bandwidth $B = 1/2T$.

(f) The threshold levels used in the decision device are $0, \pm A, \ldots, \pm(M-2)A/2$ if M is even, and $\pm A/2, \pm 3A/2, \ldots, \pm(M-2)A/2$ if M is odd.

The average probability of symbol error in this system is defined as

$$P_e = \left(1 - \frac{1}{M}\right) \operatorname{erfc}\left(\frac{A}{2\sqrt{2}\sigma}\right),$$

where σ is the standard deviation of the noise at the input of the decision device. Demonstrate the validity of this general formula by determining P_e for the following three cases: $M = 2, 3, 4$.

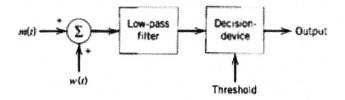

8. Answer regarding DSL.

    (a) Derive the formula for the power spectral density of a transmitted signal using the 2B1Q line code.

    (b) Plot the power spectrum of the following line codes:

- Manchester code
- Modified duobinary code
- Bipolar return-to-zero code
- 2B1Q code

Hence compare the relative merits of these line codes for their suitability in a digital subscriber loop.

9. The following figure shows the cascade connection of a linear channel and a synchronous tap-delay-line equalizer. The impulse response of the channel is denoted by $c(t)$, and that of the equalizer is denoted by $h(t)$. $h(t)$ is defined as

$$h(t) = \sum_{k=-N}^{N} w_k \delta(t - kT),$$

where T is the spacing between adjacent taps of the equalizer, and w_k are its tap-weights (coefficients). The impulse response of the cascaded system of the following figure is denoted by $p(t)$. $p(t)$ is sampled uniformly at the rate of $1/T$. To eliminate intersymbol interference, we require that the Nyquist criterion for distortionless transmission be satisfied, as shown by

$$p(nT) = \begin{cases} 1, & n = 0 \\ 0, & n \neq 0. \end{cases}$$

    (a) By imposing this condition, show that the $(2N + 1)$ tap-weights of the resulting zero-forcing equalizer satisfy the following set of $(2N + 1)$ simultaneous equations:

$$\sum_{k=-N}^{N} w_k c_{n-k} = \begin{cases} 1, & n = 0 \\ 0, & n = \pm 1, \pm 2, \ldots, \pm N, \end{cases}$$

where $c_n = c(nT)$. Hence, show that the zero-forcing equalizer is an *inverse filter* in that its transfer function is equal to the reciprocal of the transfer function of the channel.

(b) A shortcoming of the zero-forcing equalizer is *noise enhancement* that can result in poor performance in the presence of channel noise. To explore this phenomenon, consider a low-pass channel with a notch at the Nyquist frequency, that is, $H(f)$ is zero at $f = 1/2T$. Assuming that the channel noise is additive and white, show that the power spectral density of the noise at the equalizer output approaches infinity at $f = 1/2T$.

Even if the channel has no notch in its frequency response, the power spectral density of the noise at the equalizer output can assume high values. Justify the validity of this general statement.

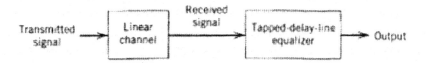

10. Some radio systems suffer from multipath distortion, which is caused by the existence of more than one propagation path between the transmitter and the receiver. Consider a channel, the output of which, in response to a signal $s(t)$, is defined as (in the absence of noise)

$$(x)t = a_1 s(t - t_{01}) + a_2 s(t - t_{02}),$$

where a_1 and a_2 are constants, and t_{01} and t_{02} represent transmission delays. It is proposed to use the three-tap delay-line-filter of the following figure to equalize the multipath distortion produced by this channel.

(a) Evaluate the transfer function of the channel.

(b) Evaluate the parameters of the tapped-delay-line filter in terms of a_1, a_2, t_{01}, and t_{02}, assuming that $a_2 \ll a_1$ and $t_{02} > t_{01}$.

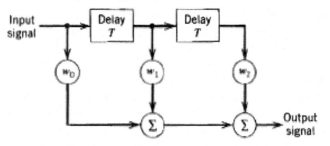

11. In Section 4.11, we studied the eye diagram of a quaternary ($M = 4$) PAM baseband transmission system under both noisy and band-limited conditions. In that experiment, the channel was assumed to be linear. In a strictly linear system with truly random data, all the eye openings would be identical. In practice, however, it is often possible to discern asymmetries in the eye pattern, which are caused by nonlinearities in the communication channel. In this experiment, we study the effect of a non-linear channel on the openings of an eye pattern. Specifically, we repeat the computer experiment pertaining to the noiseless eye pattern for $M = 4$, but this time assume that the channel is nonlinear with the following input–output relation:

$$x(t) = s(t) + as^2(t),$$

where $s(t)$ is the channel input and $x(t)$ is the channel output, and a is a constant.

(a) Do the experiment for $a = 0$, 0.05, 0.1, 0.2.

(b) Hence, discuss how varying a affects the shape of the eye pattern.

12. A stationary random process has an autocorrelation function given by $Rx = \frac{A^2}{2}e^{-|\tau|}\cos 2\pi f_0 t$ and it is known that random process never exceeds 6 in magnitude. Assuming $A = 6$,

(a) How many quantization levels are required to guarantee an SQNR of at least 60 dB?

(b) Assuming that the signal is quantized to satisfy the condition of part 1 and assuming the approximate bandwidth of the signal is W, what is the minimum required bandwidth for transmission of a binary PCM signal based on this quantization scheme?

13. A signal can be modeled as a low-pass stationary process $X(t)$ whose PDF at any time t_0 is given as the following figure.

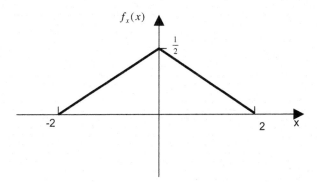

The bandwidth of this process is $5\,\text{KHz}$, and it is desired to transmit it using a PCM system.

(a) If sampling is done at the Nyquist rate and a uniform quantizer with 32 levels is employed, what is the resulting SQNR? What is the resulting bit rate?

(b) If the available bandwidth of the channel is $40\,\text{KHz}$, what is the highest achievable SQNR?

(c) If instead of sampling at the Nyquist rate we require a guard-band of at least $2\,\text{KHz}$, and the bandwidth of the channel is $40\,\text{KHz}$ again, what is the highest achievable SQNR?

14. Design an optimal compander for a source with a triangular PDF given by

$$f_x(x) = \begin{cases} x + 1, & -1 \le x \le 0 \\ -x + 1, & 0 \le x \le -1 \\ 0, & \text{otherwise.} \end{cases}$$

5

Optimal Receiver of Digital Communication Systems

From previous chapters, we recall the performance of digital communication systems to be evaluated by *bit error rate* (BER). Influenced by the development of modern probability theory and statistical decision theory, communication theory has been developed to evaluate the performance of digital communication systems and to design more effective systems meeting the needs of engineering. The first subject that we want to know is how to design an optimal receiver for a digital communication system given a signaling method being determined in advance. In other words, we shall develop a systematic way to design a mechanism of digital communication receiver that can determine whether "1" or "0" to be transmitted.

This design must be able to combat with at least two dimensions of challenges: noise from the channel (including the components used in transmitter and receiver) and the signal distortion through the channel. We may consider a phenomenon for baseband pulse transmission, after introducing pulse modulation in the previous chapter. Figure 5.1 illustrates a perfect pulse transmission as dot line drawing. The solid drawing shows pulses after transmission in the channel to result in signal *dispersion*, which can be understood by considering a pulse after non-perfect channel *filtering*filtering. Perfect channel has an impulse response a delta function, but it is not in practical situations. We can consequently observe the "overlapping" of the pulses after transmission (i.e., at the receiver decision input), which results in unwanted situation that increase the difficulty for receiver to make a right decision between "1" and "0." This unwanted signal "overlapping" is generally known as *inter-symbol interference* (ISI), while a pulse may carry information more than a bit and it is considered as a *symbol*. ISI may create

135

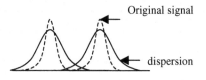

Original signal

dispersion

Fig. 5.1 Signal dispersion.

signal distortion to influence the right decision of the receiver, and may result in un-recoverable distortion at the receiver end. In this chapter, we would like to derive an optimal receiver based on a mathematical structure for digital communications.

5.1 Matched Filter

One of the best early descriptions of matched filter might be the book *Introduction to Statistical Communication Theory* authored by John Thomas in 1960s. We pretty much follow the same way in the following development. We wish to identify a linear filter that can optimize the decision of baseband pulse transmission as shown in Figure 5.2, to correctly detect the pulse transmitted over channel with embedded noise.

Suppose that the linear time-invariant filter has an impulse response of $h(t)$. The filter input (or received waveform) $x(t)$ consists of the signal pulse $g(t)$ and additive noise $w(t)$ of two-side power spectral density $N_0/2$ and is given by

$$x(t) = g(t) + w(t), \quad 0 \le t \le T, \tag{5.1}$$

where T is the observation interval and is usually the symbol period/time. The pulse binary signal may represent either "1" or "0" in a digital communication system. We wish to detect signal pulse and

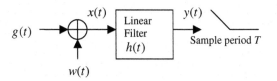

Fig. 5.2 Matched filter for a linear receiver.

make a decision about "1" or "0" based on the received waveform, under the uncertainty of noise. We consequently wish to optimize filter design in order to minimize noise effects and to optimize detection. Modern communication theory considers such a problem in a statistical manner. The resulting filter output is given by

$$y(t) = g_o(t) + n(t), \tag{5.2}$$

where $g_o(t)$ and $n(t)$ are produced by signal and noise, respectively. Recall the signal-to-noise ratio (*SNR*) to be a good index of system performance. We therefore maximize the *peak pulse SNR* as

$$\eta = \frac{|g_o(T)|^2}{E[n^2(t)]}. \tag{5.3}$$

In other words, we wish to design a linear filter to optimize its output *SNR*. Let $g(t) \leftrightarrow G(f), h(t) \leftrightarrow H(f)$. Using the inverse Fourier transform, we have

$$g_o(t) = \int_{-\infty}^{\infty} H(f)G(f)\exp(j2\pi ft)\,df. \tag{5.4}$$

When the filter output is sampled at time $t = T$, we have

$$|g_o(T)|^2 = \left| \int_{-\infty}^{\infty} H(f)\,G(f)\exp(j2\pi fT)\,df \right|^2. \tag{5.5}$$

Using the property in Chapter 2, the power spectral density of the output noise $n(t)$ is given by

$$S_N(f) = \frac{N_0}{2}|H(f)|^2. \tag{5.6}$$

The average power of the output noise $n(t)$ is expressed as

$$E[n^2(t)] = \int_{-\infty}^{\infty} S_N(f)\,df$$

$$= \frac{N_0}{2}\int_{-\infty}^{\infty} |H(f)|^2\,df. \tag{5.7}$$

We are ready to re-write the peak pulse *SNR* as

$$\eta = \frac{\left| \int_{-\infty}^{\infty} H(f)G(f)\exp(j2\pi fT)\,df \right|^2}{\frac{N_0}{2}\int_{-\infty}^{\infty} |H(f)|^2\,df}. \tag{5.8}$$

Lemma 5.1 (Cauchy's Inequality).

$$\left(\sum x_i y_i\right)^2 \leq \left(\sum x_i^2\right)\left(\sum y_i^2\right). \tag{5.9}$$

Lemma 5.2 (Schwartz Inequality). If $\int_{-\infty}^{\infty} |\phi_1(x)|^2 \, dx < \infty$ and $\int_{-\infty}^{\infty} |\phi_2(x)|^2 \, dx < \infty$, then

$$\left|\int \phi_1(x)\phi_2(x)dx\right|^2 \leq \int |\phi_1(x)|^2 dx \int |\phi_2(x)|^2 dx. \tag{5.10}$$

Equality holds when

$$\phi_1(x) = k_1 \phi_2^*(x) \tag{5.11}$$

and k_1 is a constant.

Applying Schwartz inequality, we obtain

$$\left|\int_{-\infty}^{\infty} H(f)G(f)\exp(j2\pi fT)\,df\right|^2 \leq \int_{-\infty}^{\infty} |H(f)|^2 \, df \int_{-\infty}^{\infty} |G(f)|^2 \, df. \tag{5.12}$$

We can further obtain

$$\eta \leq \frac{2}{N_0} \int_{-\infty}^{\infty} |G(f)|^2 \, df. \tag{5.13}$$

It suggests that η reaches maximum when $H(f)$ is selected to hold the equality. Thus we obtain (k is a scaling constant)

$$\eta_{\max} = \frac{2}{N_0} \int_{-\infty}^{\infty} |G(f)|^2 \, df; \tag{5.14}$$

$$H_{\text{opt}}(f) = kG^*(f)\exp(-j2\pi fT). \tag{5.15}$$

It is interesting to note that the optimal filter is in fact the complex conjugate of the original input signal except the factor $k\exp(-j2\pi fT)$. Taking the inverse Fourier transform, the impulse response of the optimal filter is given by

$$h_{\text{opt}}(t) = k \int_{-\infty}^{\infty} G^*(f)\exp[-j2\pi f(T-t)]\,df. \tag{5.16}$$

For a real signal $g(t)$, we have $G^*(f) = G(-f)$ and can express

$$h_{opt}(t) = k \int_{-\infty}^{\infty} G(-f) \exp[-j2\pi f(T-t)] \, df$$

$$= k \int_{-\infty}^{\infty} G(f) \exp[j2\pi f(T-t)] \, df$$

$$= k g(T-t). \tag{5.17}$$

(5.17) implies us that the impulse response of the optimal filter is a time-reversed and delayed version of the input signal $g(t)$, except the scaling factor k. Or, we may interpret that $h_{opt}(t)$ is matched to $g(t)$ but in a reverse direction. That is the reason why we call such a filter as a *matched filter*. Also note that we assume only noise to be white, and no further assumption about its statistics. In the frequency domain, ignoring the delay factor, the matched filter is the complex conjugate of $G(f)$.

Proposition 5.3. The matched filter optimizes the output *SNR* under the structure of Figure 5.2 having time-domain and frequency-domain response as

$$h_{opt}(t) = k g(T-t)$$
$$H_{opt}(f) = kG^*(f) \exp(-j2\pi fT).$$

Property 5.1. The peak *SNR* of the matched filter depends solely on the ratio of signal energy to the power spectral density of white noise at the filter input.

Proof. The Fourier transform of matched filter output $g_o(t)$ is given by

$$G_o(f) = H_{opt}(f)G(f)$$

$$= kG^*(f)G(f) \exp(-j2\pi fT)$$

$$= k|G(f)|^2 \exp(-j2\pi fT). \tag{5.18}$$

Using the inverse Fourier transform, we can derive the matched filter output at $t = T$ as

$$g_o(T) = \int_{-\infty}^{\infty} G_o(f) \exp(j2\pi fT) \, df$$

$$= k \int_{-\infty}^{\infty} |G(f)|^2 \, df.$$

According to Rayleigh's energy theorem (i.e., the total energy in time domain and frequency domain is the same),

$$E = \int_{-\infty}^{\infty} g^2(t) \, dt = \int_{-\infty}^{\infty} |G(f)|^2 \, df.$$

Then,

$$g_o(T) = kE. \tag{5.19}$$

The average output noise power is given by

$$E[n^2(t)] = \frac{k^2 N_o}{2} \int_{-\infty}^{\infty} |G(f)|^2 \, df$$

$$= k^2 N_o E/2. \tag{5.20}$$

Consequently, the peak *SNR* reaches its maximum, that is,

$$\eta_{\max} = \frac{(kE)^2}{(k^2 N_0 E/2)} = \frac{2E}{N_0}. \tag{5.21}$$

$\square$

Example 5.1. Let us consider the operation of a matched filter for a typical signaling $g(t)$ with a rectangular pulse of amplitude A and duration T. Figure 5.3 shows the waveform and operating waveform from the matched filter. If we use the integrate-and-dump circuit as shown in Figure 5.4, we can obtain the matched filter output at $t = T$ as shown in Figure 5.3(c).

We are now ready to evaluate the error rate of detection pulse modulation, which is the most important performance index of a digital communication system. Before we develop a general signal space methodology in the later section, we introduce the statistical

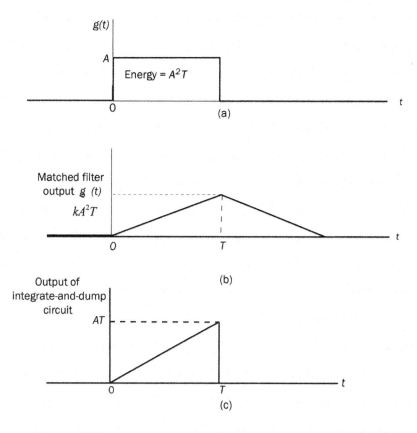

Fig. 5.3 (a) Signal. (b) Matched filter output. (c) Integrator output waveform.

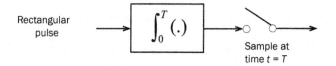

Fig. 5.4 Integrate-and-dump circuit.

decision concept to design the receiver and thus to evaluate error rate (see Figure 5.5).

Suppose, we consider a binary pulse modulation using non-return-to-zero signaling format with bit period T_b and amplitude A. The channel is embedded an additive white Gaussian noise (AWGN) with zero mean and power spectral density $N_0/2$. The received waveform can be

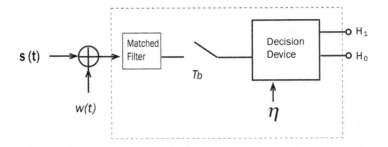

Fig. 5.5 Receiver with matched filter.

modeled as

$$x(t) = \begin{cases} +A + w(t), & \text{symbol ``1'' was sent (i.e. } H_1) \\ -A + w(t), & \text{symbol ``0'' was sent (i.e. } H_0). \end{cases} \tag{5.22}$$

The mechanism decides H_1 or H_0 based on the threshold η. There are two possible kinds of error is shown in Figure 5.5:

- Symbol "1" is determined when "0" is actually transmitted. It is the error of the first kind. We may also call such an error as *false alarm*.
- Symbol "0" is determined when "1" is actually transmitted. It is the error of the second kind. We may also call such an error as *missing*.

It is obvious for the importance of the selection of threshold. For (5.22), if we consider *equally probable* signaling (that is, *a priori* probabilities for "1" and "0" are equal or $1/2$), we may observe from Figure 5.6 to see the conditional distributions for H_1 or H_0. The threshold can be easily decided as $\eta = 0$ by symmetry.

Before we calculate the error rate (or the probability of error), we introduce the Gaussian tail function as follows:

$$Q(x) = \int_x^\infty \frac{1}{\sqrt{2\pi}} \exp\left(-\frac{x^2}{2}\right) dx. \tag{5.23}$$

Suppose that we first consider the case "0" is sent. The received signal is thus given by

$$x(t) = -A + w(t), \quad \leq t \leq T_b. \tag{5.24}$$

The sample at $t = T_b$ of a matched filter output is given by

$$y = \int_0^{T_b} x(t)\, dt$$

$$= -A + \frac{1}{T_b} \int_0^{T_b} w(t)\, dt. \qquad (5.25)$$

It represents a sample value from a random variable Y. Since $w(t)$ is white and Gaussian, we can characterize Y as

- Y is Gaussian distributed with mean $-A$.
- The variance of Y is given by

$$\sigma_Y^2 = E[(Y + A)^2]$$

$$= \frac{1}{T_b^2} E\left[\int_0^{T_b} \int_0^{T_b} w(t)w(u)\, dt\, du \right]$$

$$= \frac{1}{T_b^2} \int_0^{T_b} \int_0^{T_b} E[w(t)w(u)]\, dt\, du$$

$$= \frac{1}{T_b^2} \int_0^{T_b} \int_0^{T_b} R_W(t, u)\, dt\, du. \qquad (5.26)$$

where $R_W(t, u)$ is the autocorrelation of $w(t)$ that is white with power spectral density $N_0/2$ and is given by

$$R_W(t, u) = \frac{N_0}{2} \delta(t - u). \qquad (5.27)$$

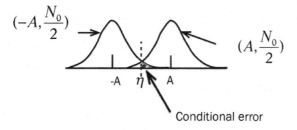

Fig. 5.6 Conditional distributions of H_1 or H_0.

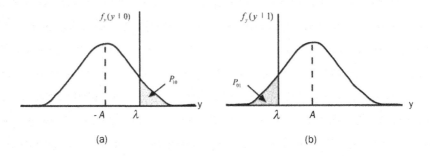

Fig. 5.7 Graphic conditional distributions.

Consequently,

$$\sigma_Y^2 = \frac{1}{T_b^2} \int_0^{T_b} \int_0^{T_b} \frac{N_0}{2} \delta(t - u) \, dt \, du$$

$$= \frac{N_0}{2T_b}. \tag{5.28}$$

The conditional probability density function of Y given symbol "0" being sent is therefore

$$f_Y(y|0) = \frac{1}{\sqrt{\pi N_0/T_b}} \exp\left(-\frac{(y + A)^2}{N_0/T_b}\right). \tag{5.29}$$

We denote the conditional probability of error given "0" being sent as p_{10} as shown in Figure 5.7.

$$p_{10} = P(y > \eta | \text{symbol 0 was sent})$$

$$= \int_\eta^\infty f_Y(y|0) \, dy$$

$$= \frac{1}{\sqrt{\pi N_0/T_b}} \int_\eta^\infty \exp\left(-\frac{(y + A)^2}{N_0/T_b}\right) dy$$

$$= Q\left(\frac{A + \eta}{\sqrt{N_0/2T_b}}\right). \tag{5.30}$$

Similarly, using conditional distribution, we have

$$f_Y(y|1) = \frac{1}{\sqrt{\pi N_0/T_b}} \exp\left(-\frac{(y - A)^2}{N_0/T_b}\right). \tag{5.31}$$

We can obtain the conditional probability error p_{01} as

$$p_{01} = P(y < \eta | \text{symbol 1 was sent})$$

$$= \int_{-\infty}^{\eta} f_Y(y|1)\, dy$$

$$= \frac{1}{\sqrt{\pi N_0/T_b}} \int_{-\infty}^{\eta} \exp\left(-\frac{(y-A)^2}{N_0/T_b}\right) dy$$

$$= Q\left(\frac{A - \eta}{\sqrt{N_0/2T_b}}\right). \tag{5.32}$$

To evaluate the performance, we can derive the *average probability of (symbol) error*, P_e, using conditional probability of error weighted by *a priori* probability distribution of "1" and "0":

$$P_e = p_0 p_{10} + p_1 p_{01}$$

$$= p_0 Q\left(\frac{A + \eta}{\sqrt{N_0/2T_b}}\right) + p_1 Q\left(\frac{A - \eta}{\sqrt{N_0/2T_b}}\right). \tag{5.33}$$

Lemma 5.4 (Leibniz's Rule). Considering the integral $\int_{a(u)}^{b(u)} f(z,u)\, dz$, the derivative of this integral with respect to u is given by

$$\frac{d}{du} \int_{a(u)}^{b(u)} f(z,u)\, dz$$

$$= f(b(u), u)\frac{db(u)}{du} - f(a(u), u)\frac{da(u)}{du} + \int_{a(u)}^{b(u)} \frac{\partial f(z,u)}{\partial u}\, dz.$$

Differentiating (5.69) with respect to η using Lemma 5.1, the optimal threshold can be derived as

$$\eta_{\text{opt}} = \frac{N_0/2}{2AT_b} \log\left(\frac{p_0}{p_1}\right). \tag{5.34}$$

For the most interesting case of equally probable signaling $p_1 = p_0 = \frac{1}{2}$, we can have $\eta_{\text{opt}} = 0$, and thus $p_{01} = p_{10}$. The resulting average probability of error is given by

$$P_e = Q\left(\frac{A}{\sqrt{N_0/2T_b}}\right). \tag{5.35}$$

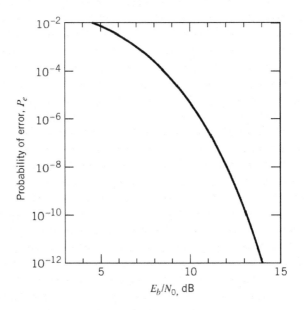

Fig. 5.8 Probability of error for binary equally probable signaling.

Recall that the transmitted energy per bit/symbol is defined as

$$E_b = A^2 T_b. \tag{5.36}$$

The average probability of error is therefore a sole function of E_b/N_0 (the ratio of transmitted energy per bit to noise spectral density). A typical plot of error rate is shown in Figure 5.8. Note that the figure is plotted in logarithm scale and it suggests a very fast decreasing of error rate (more than exponential):

$$P_e = Q\left(\sqrt{\frac{2E_b}{N_0}}\right). \tag{5.37}$$

Example 5.2 (Four-Level Pulse Amplitude Modulation (PAM)). In the above discussion, we consider binary pulse amplitude modulation (PAM) signaling. We may extend into four-level PAM with signal constellations at $-3A, -A, +A, +3A$ of equally probable distribution. Note that each pulse (symbol) can carry 2-bit information in this case. The decision thresholds are set at $-2A, 0, 2A$. Assuming normalized symbol duration and using the same receiver structure with the above decision thresholds, the symbol error rate, p_e, (not BER) is given by

$$p_e = \left(\frac{1}{4} + \frac{1}{4}\right) Q\left(\frac{A}{\sqrt{N_0/2}}\right) + \left(\frac{1}{4} + \frac{1}{4}\right) 2Q\left(\frac{A}{\sqrt{N_0/2}}\right)$$

$$= \frac{3}{2} Q\left(\frac{A}{\sqrt{N_0/2}}\right).$$

Then, we may find the average energy per bit as $E_b = 5A^2$. The average symbol probability of error (symbol error rate) is therefore given by

$$p_e = \frac{3}{2} Q\left(\sqrt{\frac{E_b/5}{N_0/2}}\right).$$

In order to achieve the same error rate, four-level PAM uses much more power for transmission than binary PAM.

5.2   Linear Receiver

As we know from the previous description, signal dispersion can create ISI in baseband pulse transmission systems. Among so many pulse modulations, we can use PAM as the representative to study such a problem. We may consider a generic binary baseband PAM system as shown in Figure 5.9. The input binary data sequence $\{b_k\}$ consists of

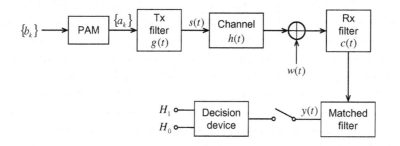

Fig. 5.9 Binary baseband data transmission system.

symbols "1" and "0" with symbol/bit duration T_b. The PAM modulator translates this binary sequence into a sequence of pulses whose amplitude is in polar form is:

$$a_k = \begin{cases} +1 & \text{if symbol } b_k \text{ is } 1 \\ -1 & \text{if symbol } b_k \text{ is } 0. \end{cases} \tag{5.38}$$

The transmit filter of an impulse response $g(t)$ conducting *waveform shaping* delivers the following transmitted signal waveform:

$$s(t) = \sum_k a_k g(t - kT_b). \tag{5.39}$$

The channel is also modeled as a filter of impulse response $h(t)$, with embedded additive noise $w(t)$ that is usually modeled as AWGN. The noisy signal after the channel $x(t)$ is then passed through a receive filter of impulse response $c(t)$. The resulting output is fed into the matched filter whose output $y(t)$ is synchronously sampled with the transmitter. Such a *synchronization* is usually achieved by extracting timing information from the received waveform. The decision device has a threshold η to determine whether "1" or "0" is received. The receive filter output is thus given by

$$y(t) = \mu \sum_k a_k p(t - kT_b - t_0) + n(t), \tag{5.40}$$

where μ is a scaling factor and t_0 can be considered as zero (perfect synchronization) without loss of generality. The scaled pulse $\mu p(t)$ with normalization $p(0) = 1$ can be expressed in both time domain and frequency domain as follows:

$$\mu p(t) = g(t) * h(t) * c(t) \tag{5.41}$$
$$\mu P(f) = G(f)H(f)C(f), \tag{5.42}$$

where $p(t) \leftrightarrow P(f)$, $h(t) \leftrightarrow H(f)$, $g(t) \leftrightarrow G(f)$, $c(t) \leftrightarrow C(f)$. $n(t)$ is the noise term from the receive filter output. The receive filter output

$y(t)$ is sampled at $t_i = iT_b$ to yield

$$y(t_i) = \mu \sum_{k=-\infty}^{\infty} a_k p[(i-k)T_b] + n(t_i)$$

$$= \mu a_i + \mu \sum_{\substack{k=-\infty \\ k \neq i}}^{\infty} a_k p[(i-k)T_b] + n(t_i). \qquad (5.43)$$

The first term is contributed from the desired ith transmitted bit; the second term is the residual effect of all other transmitted bits before/after the sampling instant and is called ISI; the third term is obviously noise contribution. Without noise and ISI, the desired output is $y(t_i) = \mu a_i$.

Since the frequency response of the channel and the transmitted pulse shaping are typically specified, the response of transmit and receive filters is determined from the reconstruction of data sequence $\{b_k\}$. The receiver has to estimate (or to decode) $\{a_k\}$ from output $y(t)$. Such a decoding requires the weighted pulse contribution free from ISI, which suggests the control of overall pulse $p(t)$ as

$$p(iT_b - kT_b) = \begin{cases} 1, & i=k \\ 0, & i \neq k. \end{cases} \qquad (5.44)$$

If $p(t)$ satisfies (5.97), then $y(t_i) = \mu a_i$, $\forall i$, to ensure perfect reception in absence of noise. To transform (5.44) into frequency domain, recalling sampling process, we can have

$$P_\delta(f) = R_b \sum_{n=-\infty}^{\infty} P(f - nR_b), \qquad (5.45)$$

where $R_b = 1/T_b$ is the bit (symbol) rate; $P_\delta(f)$ is the Fourier transform of an infinite periodic sequence of delta functions of period T_b. That is,

$$P_\delta(f) = \int_{-\infty}^{\infty} \sum_{m=-\infty}^{\infty} [p(mT_b)\delta(t - mT_b)] \exp(-j2\pi f t) \, dt. \qquad (5.46)$$

Imposing (5.44) into (5.46), we obtain

$$P_\delta(f) = \int_{-\infty}^{\infty} p(0)\delta(t) \exp(-j2\pi f t) \, dt$$

$$= p(0). \qquad (5.47)$$

By normalization ($p(0) = 1$), we can reach the *Nyquist criterion* for distortionless baseband transmission in the absence of noise.

Theorem 5.5 (Nyquist Criterion). The condition of zero observed ISI is satisfied if

$$\sum_{n=-\infty}^{\infty} P(f - nR_b) = T_b. \tag{5.48}$$

The folded frequency function $P(f)$ eliminates ISI for samples taken at interval T_b provided (5.48) is satisfied, note that the folded spectrum in (5.48) and only for samples taken at interval T_b.

The simplest way to facilitate (5.48) is to specify frequency spectrum in the form of a rectangular function

$$P(f) = \begin{cases} \frac{1}{2W}, & -W < f < W \\ 0, & |f| > W \end{cases}$$

$$= \frac{1}{2W} \text{rect}\left(\frac{f}{2W}\right). \tag{5.49}$$

The system bandwidth is given by

$$W = \frac{R_b}{2} = \frac{1}{2T_b}. \tag{5.50}$$

The desired pulse function without ISI is thus expressed as

$$p(t) = \frac{\sin(2\pi W t)}{2\pi W t}$$

$$= \text{sinc}(2Wt). \tag{5.51}$$

$R_b = 2W$ is called *Nyquist rate* and W is the Nyquist bandwidth. (5.49) and (5.51) describe an *ideal Nyquist channel*. There are two challenges in practical system design:

1. It is physically unrealizable due to the instantaneous transitions at $\pm W$
2. (5.51) decays as $\frac{1}{|t|}$, which is sensitive to sampling times.

To evaluate the effect of timing error (denoted by Δt), in the absence of noise, we have

$$y(\Delta t) = \mu \sum_k a_k p(\Delta t - kT_b)$$

$$= \mu \sum_k a_k \frac{\sin[2\pi W(\Delta t - kT_b)]}{2\pi W(\Delta t - kT_b)}. \qquad (5.52)$$

Since $2WT_b = 1$, we have

$$y(\Delta t) = \mu a_0 \operatorname{sinc}(2W\,\Delta t) + \frac{\mu \sin(2\pi W\,\Delta t)}{\pi} \sum_{\substack{k \\ k \neq 0}} \frac{(-1)^k a_k}{(2W\,\Delta t - k)}. \qquad (5.53)$$

It is the desired signal plus ISI caused by timing error.

When we are conducting communication system design, zero-forcing (ZF) technique can be used by meaning to force zero ISI at the sampling instants typically using Nyquist criterion.

5.3   Optimal Linear Receiver

Up to this point, we treat noise and ISI separately. However, in practical applications, we have to deal both at the same time. A common approach to design a linear receiver consists of a ZF equalizer followed by a decision device. The objective of an equalizer is to keep ISI forced to zero at sampling time instants. The *ZF equalizer* is relatively easy to design when ignoring the channel noise, at the price of *noise enhancement*. Another commonly used approach is based on the *minimum mean-squared error* (MMSE) criterion.

Again, considering the scenario of Figure 5.9, the received waveform after receive filter has the following response:

$$y(t) = \int_{-\infty}^{\infty} c(\tau)x(t - \tau)d\tau. \qquad (5.54)$$

The channel output is thus given by

$$x(t) = \sum_k a_k q(t - kT_b) + w(t), \qquad (5.55)$$

where a_k is the symbol transmitted at time $t = kT_b$ and $w(t)$ is the additive channel noise. $q(t) = g(t) * h(t)$, which can be considered as the pulse shape function after the channel. We can re-write (5.54) and sample the resulting output waveform $y(t)$ at $t = iT_b$.

$$y(iT_b) = \xi_i + n_i, \tag{5.56}$$

where ξ_i is the contributed from the signal and n_i is the contributed from noise and are expressed as

$$\xi_i = \sum_k a_k \int_{-\infty}^{\infty} c(\tau)q(iT_b - kT_b - \tau)d\tau \tag{5.57}$$

$$n_i = \int_{-\infty}^{\infty} c(\tau)w(iT_b - \tau)d\tau. \tag{5.58}$$

We define that the error signal as the difference between the estimated signal and the true signal to result in

$$e_i = y(iT_b) - a_i$$
$$= \xi_i + n_i - a_i. \tag{5.59}$$

The mean squared-error (MSE) is defined as

$$J = \frac{1}{2}E\left[e_i^2\right], \tag{5.60}$$

where $1/2$ is introduced as a scaling factor. Substituting (5.59) into (5.60) gives

$$J = \frac{1}{2}E[\xi_i^2] + \frac{1}{2}E\left[n_i^2\right] + \frac{1}{2}E\left[a_i^2\right] + E\left[\xi_i n_i\right] - E[n_i a_i] - E\left[\xi_i a_i\right]. \tag{5.61}$$

There are six terms in the above equation, and we evaluate one by one.

- In a stationary environment, $E[\xi_i^2]$ is independent of sampling time instants, and thus

$$E[\xi_i^2] = \sum_l \sum_k E[a_l a_k] \int_{-\infty}^{\infty} \int_{-\infty}^{\infty} c(\tau_1) c(\tau_2)$$
$$\times q(lT_b - \tau_1) q(kT_b - \tau_2) d\tau_1 d\tau_2.$$

We first assume binary symbol $a_k = \pm 1$, and the transmitted symbols are statistically independent. Then,

$$E[a_l a_k] = \begin{cases} 1 & \text{for } l = k \\ 0 & \text{otherwise.} \end{cases} \tag{5.62}$$

and

$$E[\xi_i^2] = \int_{-\infty}^{\infty} \int_{-\infty}^{\infty} R_q(\tau_1, \tau_2) c(\tau_1) c(\tau_2) \, d\tau_1 \, d\tau_2, \tag{5.63}$$

where the temporal autocorrelation function of the sequence $\{q(kT_b)\}$ is given by

$$R_q(\tau_1, \tau_2) = \sum_k q(kT_b - \tau_1) q(kT_b - \tau_2) \tag{5.64}$$

and the stationary sequence means

$$R_q(\tau_1, \tau_2) = R_q(\tau_2 - \tau_1) = R_q(\tau_1 - \tau_2).$$

- The mean-squared noise term is given by

$$E[n_i^2] = \int_{-\infty}^{\infty} \int_{-\infty}^{\infty} c(\tau_1) c(\tau_2) \, E[w(iT_b - \tau_1) w(iT_b - \tau_2)] \, d\tau_1 \, d\tau_2$$

$$= \int_{-\infty}^{\infty} \int_{-\infty}^{\infty} c(\tau_1) c(\tau_2) \, R_W(\tau_2 - \tau_1) \, d\tau_1 \, d\tau_2. \tag{5.65}$$

If we assume $w(t)$ to be white with power spectral density $N_0/2$, the ensemble-averaged autocorrelation function of channel noise $w(t)$ is given by

$$R_W(\tau_2 - \tau_1) = \frac{N_0}{2} \delta(\tau_2 - \tau_1). \tag{5.66}$$

Therefore,

$$E[n_i^2] = \frac{N_0}{2} \int_{-\infty}^{\infty} \int_{-\infty}^{\infty} c(\tau_1) c(\tau_2) \, \delta(\tau_2 - \tau_1) \, d\tau_1 \, d\tau_2. \tag{5.67}$$

- Since $a_i = \pm 1$, we can easily obtain

$$E[a_i^2] = 1 \quad \text{for all } i. \tag{5.68}$$

- Since ξ_i and n_i are independent, and n_i has zero mean due to channel noise $w(t)$ zero mean, we have

$$E[\xi_i n_i] = 0 \quad \text{for all } i. \tag{5.69}$$

- For similar reasons, we have

$$E[n_i a_i] = 0 \quad \text{for all } i. \tag{5.70}$$

- Using (5.57), we have

$$E[\xi_i a_i] = \sum_k E[a_k a_i] \int_{-\infty}^{\infty} c(\tau) q(iT_b - kT_b - \tau) d\tau. \tag{5.71}$$

Similar to earlier independence assumption for transmitted symbols of (5.52), we reduce (5.71) to

$$\left(S_q(f) + \frac{N_0}{2} \right) C(f) = Q^*(f). \tag{5.72}$$

The above derivations result in MSE J for binary data transmission system as

$$J = \frac{1}{2} + \frac{1}{2} \int_{-\infty}^{\infty} \int_{-\infty}^{\infty} \left(R_q(t - \tau) + \frac{N_0}{2} \delta(t - \tau) \right) c(t) c(\tau) \, dt \, d\tau$$

$$- \int_{-\infty}^{\infty} c(t) q(-t) \, dt. \tag{5.73}$$

To design the receive filter $c(t)$ as shown in Figure 5.9, we differentiate (5.73) with respect to impulse response of the receive filter $c(t)$ and set to zero (necessary condition of optimality). We thus have

$$\int_{-\infty}^{\infty} \left(R_q(t - \tau) + \frac{N_0}{2} \delta(t - \tau) \right) c(\tau) \, d\tau = q(-t). \tag{5.74}$$

On the basis of (5.74), we can identify $c(t)$ of the equalizer optimized in MSE sense, and such an equalizer is known as *MMSE equalizer*. Taking the Fourier transform $(c(t) \leftrightarrow C(f),\ q(t) \leftrightarrow Q(f),\ R_q(\tau) \leftrightarrow S_q(f))$, we obtain

$$\left(S_q(f) + \frac{N_0}{2} \right) C(f) = Q^*(f). \tag{5.75}$$

That is,

$$C(f) = \frac{Q^*(f)}{S_q(f) + \frac{N_0}{2}}. \tag{5.76}$$

We can show that the power spectral density of $\{q(kT_b)\}$ can be expressed as

$$S_q(f) = \frac{1}{T_b} \sum_k \left| Q\left(f + \frac{k}{T_b}\right) \right|^2. \tag{5.77}$$

It suggests that the frequency response $C(f)$ of the optimum linear receiver is periodic with period $1/T_b$ (see Figure 5.10).

Proposition 5.6. Assuming linear $g(t)$ and $c(t)$, the optimum linear receiver for Figure 5.9 consists of cascading two filters:

  (a) A matched filter with impulse response $q(-t)$, while $q(t) = g(t) * h(t)$;
  (b) A transversal tap-delay-line equalizer with a frequency response of the inverse of periodic function $S_q(f) + \frac{N_0}{2}$.

To deal with noise and ISI at the same time, we introduce the concept of equalizer. Equalizer can be realized by the ZF criterion at the price of noise enhancement. We can also use MMSE criterion to strike balanced optimization or both ISI and noise. Given that the optimal receiver is synchronous (i.e., perfectly recovered time, frequency, and phase), its a cascaded filtering of matched filter, equalizer, and decision device (demodulator).

The tap-delay-line equalization/matched filter is usually implemented with delay less than a symbol (due to over-sampling). A special

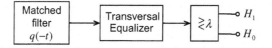

Fig. 5.10 Optimum linear receiver.

case but very common to use is spacing of adjacent taps at $T/2$ (i.e., half symbol period), and such a structure is named as *fractionally spaced equalizer*.

5.3.1 Adaptive Equalization

Ideally, if we know the channel response, a straightforward equalizer can be $\frac{1}{H(f)}$. But, $\frac{1}{H(f)}$ may not be realizable. ZF and MMSE provide general design approaches. Unfortunately, the receiver may not know the channel characteristics. Adaptive equalizer invented by Robert Lucky was introduced and has been applied even to modern digital communication systems. Figure 5.11 depicts an adaptive equalizer. Prior to data transmission, a *training sequence* known by transmitter and receiver is transmitted first so that the filter coefficients can be adjusted to appropriate values for later data communication. This stage is known as the training mode. Then, it is switched to the decision-directed mode by using coefficients obtained earlier and keeps updating based on a new decision to result in error signal $e[n] = \hat{a}[n] - y[n]$ driving further update.

The adjustment of adaptive equalizer coefficients usually relies on the steep-descent algorithms, and interesting readers can find in the literatures about various algorithms and their behaviors.

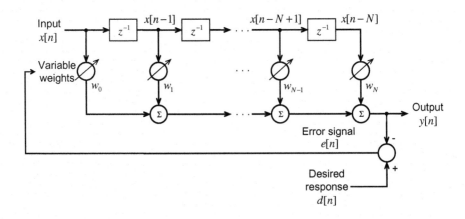

Fig. 5.11 Adaptive equalizer.

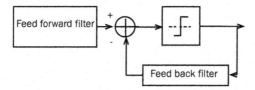

Fig. 5.12 Decision feedback equalizer.

A popular structure of adaptive filter is *decision feedback equalizer* (DFE) (DFE) as shown in Figure 5.12. The idea is based on

$$y[n] = \sum_k h[k]x[n-k]$$

$$= h[0]x[n] + \sum_{k<0} h[k]x[n-k] + \sum_{k>0} h[k]x[n-k]. \quad (5.78)$$

The second term is due to the *precursors* of the channel impulse response and the third term is due to *postcursors*. DFE is widely used in digital communication systems.

When we observe the ISI from the instrument, we may overlap waveform to see the so-called *eye pattern* as shown in Figure 5.13. The widen eye patterns mean little ISI.

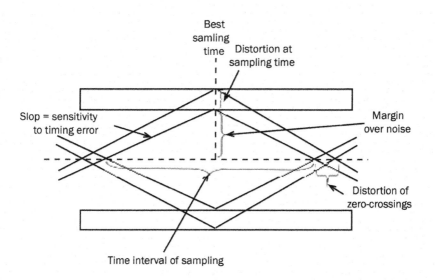

Fig. 5.13 Eye pattern.

5.3.2   Line Coding

In addition to receive filter, transmit filter design is another important subject. We generally consider as waveform shaping. The first part is because no ideal Nquist channel exists, and realistic signaling has an excess bandwidth as $B_T = W(1 + \alpha)$, where α is the *roll-off factor* to indicate excess bandwidth over the ideal Nyquist bandwidth.

Up to this point, we treat ISI as something undesirable. However, a different point of view is to control the ISI to achieve signaling at Nyquist rate. Such a methodology is called *correlative-level coding* or *line coding*, which results in *partial response* signaling. Partial response signaling implies the signaling function over one symbol period, as a counterpart of full response signaling in typical cases. It is common to consider partial response signaling in cable communication and memory storage channels (see Figure 5.14).

The basic idea of correlative coding can be illustrated by the *duo-binary signaling* which is also known as *class I partial response*class I partial response. Consider a binary input sequence $\{b_k\}$ consisting of uncorrelated binary symbols "1" and "0" with symbol duration T_b. Binary PAM modulator gives

$$a_k = \begin{cases} +1 & \text{if } b_k = 1 \\ -1 & \text{if } b_k = 0. \end{cases} \tag{5.79}$$

The sequence then is applied to *duo-binary encoder*, which converts to a three-level output, say -2, 0, and $+2$. Figure 5.15 shows the entire duo-binary signaling mechanism. The encoder output is thus given by

$$c_k = a_k + a_{k-1}. \tag{5.80}$$

Partial response is clear in this case as c_k cannot be determined solely by a_k. The frequency response of duo-binary signaling can be found in Figure 5.16 and its merits can be easily observed at the nulls of $\pm 1/2T_b$. However, its drawback is also obvious due to its memory. If there exists any error, then it can create further errors at later time, which is known as *error propagation*.

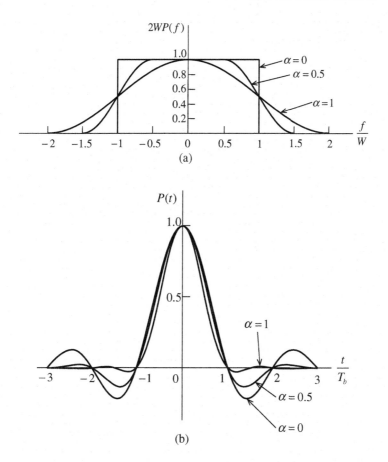

Fig. 5.14 Waveforms for different rolloff factors.

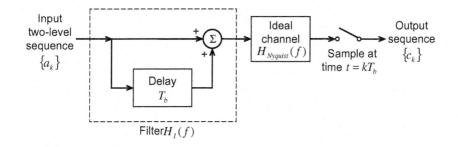

Fig. 5.15 Duo-binary signaling.

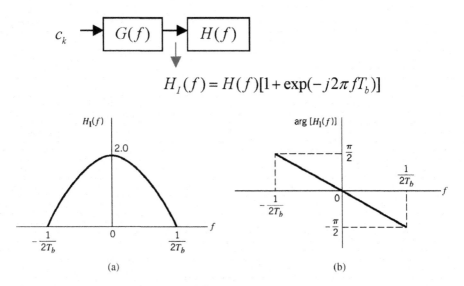

$$H_I(f) = H(f)[1 + \exp(-j2\pi fT_b)]$$

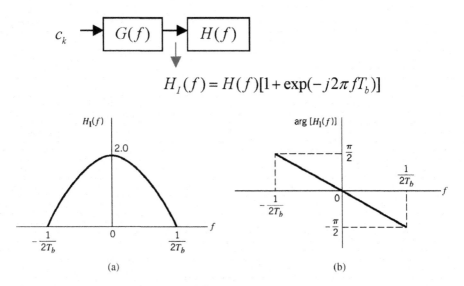

(a)   (b)

Fig. 5.16 Frequency response.

The frequency response can be derived as $H_I(f) = 2\cos(\pi fT_b)\exp(-j\pi fT_b)$, which suggests the maximum at DC term and implies waste of energy.

To avoid the error propagation, a brilliant idea named as *precoding* is used as shown in Figure 5.17, while

$$d_k = b_k \oplus d_{k-1}. \tag{5.81}$$

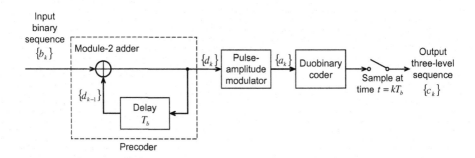

Fig. 5.17 Precoding.

5.3.3 *M*-ary PAM and Applications to Telecommunications

Baseband *M*-ary PAM:

More applications of *M*-ary PAM can be found in the following communication systems:

- Another variation of PAM is pulse position modulation (PPM) that is widely used in optical fiber transmission;
- Subscriber Line (SL): plug-in telephone line;
- DSL: transmit digital signal, e.g., modem;
- xDSL: replace the interface at end points, using the same SL in between.

5.4 Signal Space Analysis

We have described the optimal design of a receiver for digital communication system based on pulse transmission. We also conducted rather intuitive error rate analysis. In what follows, we shall develop a general treatment of communication theory for digital communications.

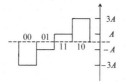

Fig. 5.18 *M*-ary PAM ($M = 4$).

To deal with the uncertainty, we usually model the phenomenon through statistics, probability, and stochastic process. Modern communication theory is based on the statistical decision theory.

On the basis of Figure 5.19, we summarize the following assumptions:

- Assuming equally probable inputs, we have

$$p_i = P(m_i)$$
$$= \frac{1}{M} \quad \text{for } i = 1, 2, \ldots, M; \tag{5.82}$$

Fig. 5.19 Block diagram of a typical digital communication system.

- Assuming equal energy for (full-response) signaling, we have

$$\int_0^T s_i^2(t)dt = E_i = E; \tag{5.83}$$

- Assuming a linear time-invariant LTI channel filter;
- Assuming AWGN.

The received signal is given by

$$x(t) = s_i(t) + w(t), \quad \begin{cases} 0 \le t \le T \\ i = 1, 2, \dots, M. \end{cases} \tag{5.84}$$

The performance measure, average probability of error, is given by

$$P_e = \sum_{i=1}^{M} p_i P(\hat{m} \ne m_i | m_i). \tag{5.85}$$

To develop a general theoretical framework, we use orthonormal function expansions

$$s_i(t) = \sum_{j=1}^{N} s_{ij} \phi_j(t), \quad \begin{cases} 0 \le t \le T \\ i = 1, 2, \dots, M. \end{cases} \tag{5.86}$$

$$s_{ij} = \int_0^T s_i(t)\,\phi_j(t)\,dt, \quad \begin{cases} i = 1, 2, \dots, M \\ j = 1, 2, \dots, N. \end{cases} \tag{5.87}$$

$$\int_0^T \phi_i(t)\,\phi_j(t)\,dt = \delta_{ij} = \begin{cases} 1 & \text{if } i = j \\ 0 & \text{if } i \ne j. \end{cases} \tag{5.88}$$

We may treat the N-dimensional vector $\mathbf{s}_i^T = (s_{i1}, s_{i2}, \dots, s_{iN})$ and it corresponds to $s_i(t)$, for $i = 1, 2, \dots, M$, while T denotes transpose.

Or, we may represent it as

$$\mathbf{s}_i = \begin{bmatrix} s_{i1} \\ s_{i2} \\ \vdots \\ s_{iN} \end{bmatrix}, \quad i = 1, 2, \ldots, M, \tag{5.89}$$

which is known as a signal vector. We may further visualize a set of M signal constellation points in N-dimensional space that is called *signal space*. It is easy to imagine such a picture under Euclidean space.

Under the concept of signal space, we want to inspect the behaviors of noise process using the following Lemma.

Lemma 5.7 (White Gaussian Noise Characterization). Suppose that $n(t)$ is a white Gaussian noise process with zero mean and power spectral density $N_0/2$. $\{\phi_i(t)\}_{i=1}^{\infty}$ form an orthonormal basis:

$$n_i = \int_0^T n(t)\phi_i(t)dt.$$

(a) $\{n_i\}_{i=1}^{\infty}$ are jointly Gaussian;

(b) $\{n_i\}_{i=1}^{\infty}$ are mutually uncorrelated and thus mutually independent;

(c) $\{n_i\}_{i=1}^{\infty}$ are iid Gaussian with zero mean and spectral variance $N_0/2$.

We generally treat the signal space as an N-dimensional vector space. The squared-length (or squared-magnitude) of a signal vector $\mathbf{s}_i$ is given by

$$\|\mathbf{s}_i\|^2 = \mathbf{s}_i^{\mathrm{T}} \mathbf{s}_i$$

$$= \sum_{j=1}^{N} s_{ij}^2, \quad i = 1, 2, \ldots, M. \tag{5.90}$$

The signal energy of $s_i(t)$ with duration T is expressed as

$$E_i = \int_0^T s_i^2(t)\, dt. \tag{5.91}$$

Its equivalent form is

$$E_i = \int_0^T \left[\sum_{j=1}^N s_{ij}\phi_j(t) \right] \left[\sum_{k=1}^N s_{ik}\phi_k(t) \right] dt.$$

By interchanging the order of summation and integration, we have

$$E_i = \sum_{j=1}^N \sum_{k=1}^N s_{ij} s_{ik} \int_0^T \phi_j(t)\,\phi_k(t)\,dt. \tag{5.92}$$

Using the orthonormal property, (5.92) reduces to

$$E_i = \sum_{j=1}^N s_{ij}^2$$

$$= \|\mathbf{s}_i\|^2. \tag{5.93}$$

For a pair of signals $s_i(t)$ and $s_k(t)$,

$$\int_0^T s_i(t)\,s_k(t)\,dt = \mathbf{s}_i^{\mathsf{T}}\mathbf{s}_k. \tag{5.94}$$

It states the equivalence of inner product between time-domain representation and signal space vector form. The squared Euclidean distance d_{ik}^2 is described as

$$\|\mathbf{s}_i - \mathbf{s}_k\|^2 = \sum_{j=1}^N (s_{ij} - s_{kj})^2$$

$$= \int_0^T (s_i(t) - s_k(t))^2\,dt. \tag{5.95}$$

The angle between two signal vectors is given by

$$\cos\theta_{ik} = \frac{\mathbf{s}_i^{\mathsf{T}}\mathbf{s}_k}{\|\mathbf{s}_i\|\,\|\mathbf{s}_k\|}. \tag{5.96}$$

Example 5.3 (Schwartz Inequality). Considering a pair of signals, Schwartz inequality states

$$\left(\int_{-\infty}^{\infty} s_1(t)\,s_2(t)\,dt \right)^2 \le \left(\int_{-\infty}^{\infty} s_1^2(t)\,dt \right) \left(\int_{-\infty}^{\infty} s_2^2(t)\,dt \right) \tag{5.97}$$

with equality holds as $s_2(t) = cs_1(t)$ and c is a constant. We can prove by expressing both signals in terms of the following two orthogonal basis functions:

$$s_1(t) = s_{11}\phi_1(t) + s_{12}\phi_2(t)$$
$$s_2(t) = s_{21}\phi_1(t) + s_{22}\phi_2(t),$$

where

$$\int_{-\infty}^{\infty} \phi_i(t)\,\phi_j(t)\, dt = \delta_{ij} = \begin{cases} 1 & \text{for } j = i \\ 0 & \text{otherwise.} \end{cases}$$

The vector form is given by

$$\mathbf{s_1} = \begin{bmatrix} s_{11} \\ s_{12} \end{bmatrix}$$

$$\mathbf{s_2} = \begin{bmatrix} s_{21} \\ s_{22} \end{bmatrix}.$$

The angle is given by

$$\cos\theta = \frac{\mathbf{s_1^T s_2}}{\|\mathbf{s_1}\| \, \|\mathbf{s_2}\|}$$

$$= \frac{\int_{-\infty}^{\infty} s_1(t)\, s_2(t)\, dt}{\left(\int_{-\infty}^{\infty} s_1^2(t)\, dt\right)^{1/2} \left(\int_{-\infty}^{\infty} s_2^2(t)\, dt\right)^{1/2}}. \qquad (5.98)$$

$|\cos\theta| \leq 1$ with equality holds if and only if $\mathbf{s_2} = c\mathbf{s_1}$ and we have

$$\left| \int_{-\infty}^{\infty} s_1(t)\, s_2^*(t)\, dt \right|^2 \leq \left(\int_{-\infty}^{\infty} |s_1(t)|^2\, dt \right) \left(\int_{-\infty}^{\infty} |s_2(t)|^2\, dt \right). \qquad (5.99)$$

The above example shows the beauty of geometric representation of signals in vector space. The next challenge would be how to determine the appropriate orthonormal basis functions. From vector analysis, this can be facilitated using *Gram–Schmidt (orthogonalization) procedure.* Suppose, that we have a set of M signals denoted by $s_1(t), \ldots, s_M(t)$. We arbitrarily select $s_1(t)$ and define the first basis function as

$$\phi_1(t) = \frac{s_1(t)}{\sqrt{E_1}}, \qquad (5.100)$$

where E_1 is the energy of signal $s_1(t)$. We therefore have

$$s_1(t) = \sqrt{E_1}\phi_1(t)$$
$$= s_{11}\,\phi_1(t). \tag{5.101}$$

$\phi_1(t)$ clearly has unit energy. Next, we define coefficient s_{21} using signal $s_2(t)$ as follows:

$$s_{21} = \int_0^T s_2(t)\,\phi_1(t)\,dt. \tag{5.102}$$

Introducing a new intermediate function orthogonal to $\phi_1(t)$,

$$g_2(t) = s_2(t) - s_{21}\,\phi_1(t). \tag{5.103}$$

The second basis function can be defined as

$$\phi_2(t) = \frac{g_2(t)}{\sqrt{\int_0^T g_2^2(t)\,dt}}. \tag{5.104}$$

By further simplifying and using E_2 as the energy of $s_2(t)$, we obtain

$$\phi_2(t) = \frac{s_2(t) - s_{21}\,\phi_1(t)}{\sqrt{E_2 - s_{21}^2}}. \tag{5.105}$$

It is clear that

$$\int_0^T \phi_2^2(t)\,dt = 1$$

$$\int_0^T \phi_1(t)\,\phi_2(t)\,dt = 0.$$

That is, $\phi_1(t)$ and $\phi_2(t)$ form an orthonormal pair. Continuing the same process, we generally define

$$g_i(t) = s_i(t) - \sum_{j=1}^{i-1} s_{ij}\phi_j(t). \tag{5.106}$$

$$s_{ij} = \int_0^T s_i(t)\,\phi_j(t)\,dt, \quad j = 1,2,\ldots,i-1. \tag{5.107}$$

The set of basis functions

$$\phi_i(t) = \frac{g_i(t)}{\sqrt{\int_0^T g_i^2(t)\,dt}}, \quad i = 1,2,\ldots,N \tag{5.108}$$

form an orthonormal set. The signal $s_1(t), \ldots, s_M(t)$ form a linearly independent set for $N = M$. The special case for $N < M$ results in $g_i(t) = 0$, $i > N$, and signals are not linearly independent.

Then we are ready to convert continuous-time AWGN channel into a vector channel form for in-depth development of mathematical framework. The AWGN channel output is fed into a bank of N correlators as shown in Figure 5.20.

$$x(t) = s_i(t) + w(t), \quad \begin{cases} 0 \leq t \leq T \\ i = 1, 2, \ldots, M, \end{cases} \tag{5.109}$$

where $w(t)$ is the sampled function of a white Gaussian noise process $W(t)$ with zero mean and power spectral density $N_0/2$. The output of correlator j is the sampled value of a random variable X_j as

$$x_j = \int_0^T x(t) \, \phi_j(t) \, dt$$

$$= s_{ij} + w_j, \quad j = 1, 2, \ldots, N. \tag{5.110}$$

Contributed from signal and noise, respectively, both terms in (5.110) are also

$$s_{ij} = \int_0^T s_i(t) \, \phi_j(t) \, dt \tag{5.111}$$

$$w_j = \int_0^T w(t) \, \phi_j(t) \, dt. \tag{5.112}$$

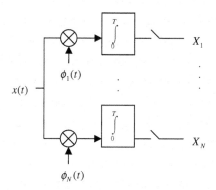

Fig. 5.20 Bank of correlators.

We now consider a new random process $X'(t)$ whose sample function is $x'(t)$ with the following relationship:

$$x'(t) = x(t) - \sum_{j=1}^{N} x_j \phi_j(t). \tag{5.113}$$

Substituting (5.109) and (5.110) into (5.111), with expansion, we have

$$x'(t) = s_i(t) + w(t) - \sum_{j=1}^{N} (s_{ij} + w_j) \phi_j(t)$$

$$= w(t) - \sum_{j=1}^{N} w_j \phi_j(t)$$

$$= w'(t). \tag{5.114}$$

The received signal is consequently

$$x(t) = \sum_{j=1}^{N} x_j \phi_j(t) + x'(t)$$

$$= \sum_{j=1}^{N} x_j \phi_j(t) + w'(t). \tag{5.115}$$

The next step is to characterize the N correlator outputs. Let $X(t)$ denote the random process whose sampled function is $x(t)$. $X(t)$ is a Gaussian process. As discribed in Chapter 2, X_j are Gaussian random variables, $j = 1, 2, \ldots, N$. To determine its mean and variance to characterize outputs,

$$\begin{aligned}
\mu_{X_j} &= E[X_j] \\
&= E[s_{ij} + W_j] \\
&= s_{ij} + E[W_j] \\
&= s_{ij}
\end{aligned} \tag{5.116}$$

and

$$\begin{aligned}
\sigma_{X_j}^2 &= \text{var}[X_j] \\
&= E\left[(X_j - s_{ij})^2\right] \\
&= E\left[W_j^2\right].
\end{aligned} \tag{5.117}$$

From (5.112),

$$W_j = \int_0^T W(t)\,\phi_j(t)\,dt.$$

By expansion,

$$\sigma_{X_j}^2 = E\left[\int_0^T W(t)\,\phi_j(t)\,dt \int_0^T W(u)\,\phi_j(u)\,du\right]$$

$$= E\left[\int_0^T \int_0^T \phi_j(t)\,\phi_j(u)\,W(t)\,W(u)\,dt\,du\right]. \qquad (5.118)$$

Interchanging expectation and integration,

$$\sigma_{X_j}^2 = \int_0^T \int_0^T \phi_j(t)\,\phi_j(u)\,E\left[W(t)\,W(u)\right]dt\,du$$

$$= \int_0^T \int_0^T \phi_j(t)\,\phi_j(u)\,R_W(t,u)\,dt\,du, \qquad (5.119)$$

where $R_W(t,u)$ is the autocorrelation function of noise process. According to the white noise property,

$$R_W(t,u) = \frac{N_0}{2}\delta(t-u). \qquad (5.120)$$

The variance is given by

$$\sigma_{X_j}^2 = \frac{N_0}{2}\int_0^T \int_0^T \phi_j(t)\,\phi_j(u)\,\delta(t-u)\,dt\,du$$

$$= \frac{N_0}{2}\int_0^T \phi_j^2(t)\,dt. \qquad (5.121)$$

According to the unit energy property of basis function, we have

$$\sigma_{X_j}^2 = \frac{N_0}{2} \quad \text{for all } j. \qquad (5.122)$$

It suggests that correlator outputs have the same variance as noise process. Furthermore, since $\{\phi(t)\}$ form an orthogonal set, X_j are mutually uncorrelated as

$$
\begin{aligned}
\operatorname{cov}[X_j X_k] &= E[(X_j - \mu_{X_j})(X_k - \mu_{X_k})] \\
&= E[(X_j - s_{ij})(X_k - s_{ik})] \\
&= E[W_j W_k] \\
&= E\left[\int_0^T W(t)\,\phi_j(t)\,dt \int_0^T W(u)\,\phi_k(u)\,du\right] \\
&= \int_0^T \int_0^T \phi_j(t)\,\phi_k(u)\,R_W(t,u)\,dt\,du \\
&= \frac{N_0}{2} \int_0^T \int_0^T \phi_j(t)\,\phi_k(u)\,\delta(t-u)\,dt\,du \\
&= \frac{N_0}{2} \int_0^T \phi_j(t)\,\phi_k(t)\,dt \\
&= 0, \quad j \neq k.
\end{aligned}
\tag{5.123}
$$

For Gaussian random variables, uncorrelated property suggests independence. Define the vector of N random variables as

$$
\mathbf{X} =
\begin{bmatrix}
X_1 \\
X_2 \\
\vdots \\
X_N
\end{bmatrix}.
\tag{5.124}
$$

The elements of (5.124) are independent Gaussian random variables with mean s_{ij} and variance $N_0/2$. The conditional probability density function of $\mathbf{X}$, given that $s_i(t)$ (or symbol m_i) is transmitted, is given by

$$
f_X(\mathbf{x}|m_i) = \prod_{j=1}^{N} f_{X_j}(x_j|m_i), \quad i = 1, 2, \ldots, M,
\tag{5.125}
$$

where the product form comes from the independence. The vector $\mathbf{x}$ is the *observation* vector. The channel satisfying (5.125) is a *memoryless* channel. Using Gaussian mean and variance, we have

$$
f_{X_j}(x_j|m_i) = \frac{1}{\sqrt{\pi N_0}} \exp\left[-\frac{1}{N_0}(x_j - s_{ij})^2\right], \quad
\begin{array}{l}
j = 1, 2, \ldots, N \\
i = 1, 2, \ldots, M.
\end{array}
\tag{5.126}
$$

The resulting conditional probability is given by

$$f_X(\mathbf{x}|m_i) = (\pi N_0)^{-N/2} \exp\left[-\frac{1}{N_0}\sum_{j=1}^{N}(x_j - s_{ij})^2\right], \quad i = 1,2,\ldots,M.$$
(5.127)

Since the noise process $W(t)$ is Gaussian with zero mean, $W'(t)$ represented by the sample function $w'(t)$ is also a zero-mean Gaussian process. In fact, $W'(t_k)$ sampling at any time t_k is independent of $\{X_j\}$. That is,

$$E[X_j W'(t_k)] = 0, \quad \begin{cases} j = 1,2,\ldots,N \\ 0 \le t_k \le T. \end{cases}$$
(5.128)

(5.127) suggests that the random variable $W'(t_k)$ is irrelevant to the decision of transmitted signal. In other words, the correlator outputs determined by the received waveform $x(t)$ are the only data useful for decision-making, and we call *sufficient statistics* for the decision.

Theorem 5.8 (Theorem of Irrelevance). For signal detection in AWGN, only the projected terms onto basis functions of signal $\{s_i(t)\}_{i=1}^{M}$ provides sufficient statistics of the decision problem. All other irrelevant information cannot help detection at all.

Corollary 5.1. AWGN channel of Figure 5.21 is equivalent to an N-dimensional vector channel described by the observation vector

$$\mathbf{x} = \mathbf{s}_i + \mathbf{w}, \quad i = 1,2,\ldots,M.$$
(5.129)

To design the optimum receiver, we have to consider the problem: Given observation vector $\mathbf{x}$, we would like to estimate the message symbol m_i generating $\mathbf{x}$. As we have obtained conditional probability density function $f_X(\mathbf{x}|m_i)$, $i = 1,2,\ldots,M$, to characterize AWGN, *likelihood function* denoted by $L(m_i)$ is introduced to design the receiver.

$$L(m_i) = f_X(\mathbf{x}|m_i), \quad i = 1,2,\ldots,M.$$
(5.130)

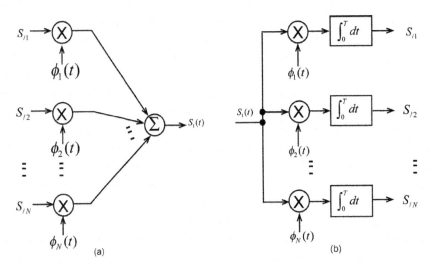

Fig. 5.21 AWGN channel by functional expansion.

In many cases, we find more convenient to use *log-likelihood function* as

$$l(m_i) = \log L(m_i), \quad i = 1, 2, \ldots, M. \tag{5.131}$$

Since the probability function is always non-negative and logarithm function is monotonically increasing, log-likelihood function is one-to-one mapping with likelihood function. For AWGN, by ignoring the constant term, we have

$$l(m_i) = -\frac{1}{N_0} \sum_{j=1}^{N} (x_j - s_{ij})^2, \quad i = 1, 2, \ldots, M. \tag{5.132}$$

Note the engineering meaning of (5.132), which is the squared Euclidean distance normalized by noise spectral density, just like E_b/N_0 or *SNR*.

5.5   Correlation Receiver

In what follows, we will use the concept of the signal space to develop *maximum likelihood* (*ML*) detection/decoding. Suppose that in a symbol duration T, one of the possible signals $s_1(t), \ldots, s_M(t)$ is with equal *a priori* probability $1/M$. By a bank of correlators, signal vector s_i is equivalent to $s_i(t)$ in such signal detection as long as $N \leq M$. The set of message points corresponding to $\{s_i(t)\}_{i=1}^{M}$ is called the *signal constellation*.

Definition 5.2 (Signal Detection Problem). Given the observation vector $\mathbf{x}$, we determine a *decision rule* (actually a mapping) to obtain estimate $\hat{m}$ of the transmitted symbol m_i, so that the probability of error can be minimized.

Given $\mathbf{x}$, we decide $\hat{m} = m_i$. The probability of error is given by

$$P_e(m_i|\mathbf{x}) = P(m_i \text{ not sent}|\mathbf{x})$$

$$= 1 - P(m_i \text{ sent}|\mathbf{x}). \tag{5.133}$$

The decision criterion is to minimize the probability of error in mapping each given observation vector $\mathbf{x}$ to a decision. The *optimum decision rule* is stated as follows:

Set $\hat{m} = m_i$ if

$$P(m_i \text{ sent } |\mathbf{x}) \geq P(m_k \text{ sent } |\mathbf{x}) \quad k = 1, 2, \ldots, M \quad \text{and} \quad k \neq i. \tag{5.134}$$

This decision rule is also known as *maximum a posteriori probability (MAP) maximum a posteriori probability (MAP)* rule. Using Baye's rule and signal *a priori* probability p_k to transmit m_k, MAP rule can be restated as

Set $\hat{m} = m_i$ if

$$\frac{p_k f_X(\mathbf{x}|m_k)}{f_X(\mathbf{x})} \quad \text{is maximum for } k = i. \tag{5.135}$$

Since $f_X(\mathbf{x})$ is independent of the transmitted signal and equal *a priori* probability p_k, the conditional probability density function having one-to-one relationship to log-likelihood function gives

Set $\hat{m} = m_i$ if

$$l(m_k) \quad \text{is maximum for } k = i. \tag{5.136}$$

(5.135) represents the *ML* rule. The mechanism to implement this rule is a *ML* decoder/detector. Note that we use equal *a priori* probability to obtain *ML* detector from MAP detector.

Leveraging the signal space concept, we can graphically explain the decision rule. Let Z denote the N-dimensional space of all possible observation vectors $\mathbf{x}$, which forms the *observation space*. The total observation space Z is partitioned into M decision regions corresponding to possible signals, and these decision regions are denoted by $Z_1, \ldots, Z_M$ (Figure 5.22). The decision rule is now

Observation vector $\mathbf{x}$ lies in region Z_i if

$$l(m_k) \text{ is maximum for } k = i. \qquad (5.137)$$

From (5.132) for AWGN, $l(m_k)$ attains its maximum when

$$\sum_{j=1}^{N} (x_j - s_{kj})^2$$

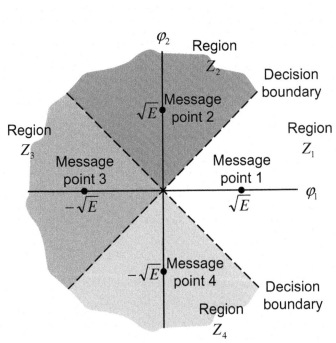

Fig. 5.22 Decision regions in signal space.

is minimized by selecting $k = i$. The ML decision rule for AWGN is

$$\text{Observation vector } \mathbf{x} \text{ lies in region } Z_i \text{ if}$$

$$\sum_{j=1}^{N}(x_j - s_{kj})^2 \quad \text{is minimum for } k = i. \tag{5.138}$$

We can use the definition of Euclidean distance to obtain

$$\sum_{j=1}^{N}(x_j - s_{kj})^2 = \|\mathbf{x} - \mathbf{s}_k\|^2. \tag{5.139}$$

The decision rule is now

$$\text{Observation vector } \mathbf{x} \text{ lies in region } Z_i \text{ if}$$
$$\text{the Euclidean distance } \|\mathbf{x} - \mathbf{s}_k\| \text{ is minimum for } k = i. \tag{5.140}$$

Theorem 5.9. The ML decision rule is simply to select the message point closest (w.r.t. Euclidean distance) to the received signal.

Straightforward algebra gives

$$\sum_{j=1}^{N}(x_j - s_{kj})^2 = \sum_{j=1}^{N} x_j^2 - 2\sum_{j=1}^{N} x_j s_{kj} + \sum_{j=1}^{N} s_{kj}^2. \tag{5.141}$$

The first term is independent of k and can be ignored. The second term is the inner product of the observation vector $\mathbf{x}$ and signal vector $\mathbf{s_k}$. The third term is the energy of the transmitted signal. Consequently, the decision rule is

$$\text{Observation vector } \mathbf{x} \text{ lies in region } Z_i \text{ if}$$

$$\sum_{j=1}^{N} x_j s_{kj} - \frac{1}{2} E_k \quad \text{is maximum for } k = i, \tag{5.142}$$

where

$$E_k = \sum_{j=1}^{N} s_{kj}^2. \tag{5.143}$$

Based on the above development, the optimum receiver for AWGN channel involves two sub-systems as shown in Figure 5.23:

(a) A bank of correlators (product-integrator) corresponding to a set of orthonormal basis functions.

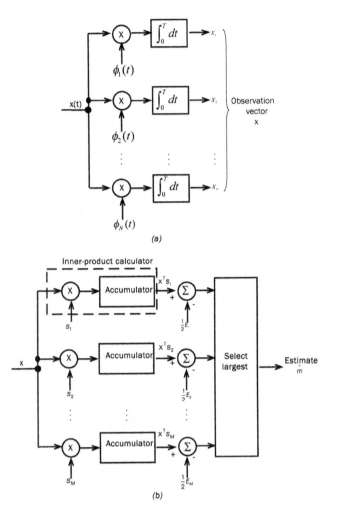

Fig. 5.23 (a) Waveform projection; (b) correlation metric calculator.

  (b) A correlation metric calculator to compute (5.141) for each
      possible signal.

Such a structure is known as *correlation receiver*.

Theorem 5.10. The correlation receiver is equivalent to the matched
filter in AWGN channels.

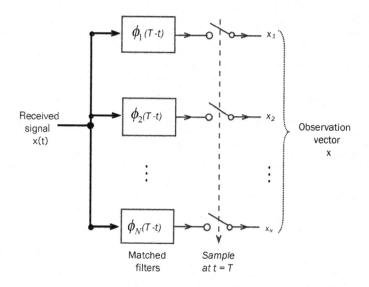

Fig. 5.24 Matched filter receiver.

Proof. With received waveform as input, the filter output is given by

$$y_j(t) = \int_{-\infty}^{\infty} x(\tau) h_j(t - \tau) \, d\tau. \qquad (5.144)$$

The matched filter is thus expressed as (Figure 5.24)

$$h_j(t) = \phi_j(T - t). \qquad (5.145)$$

The resulting filter output is given by

$$y_j(t) = \int_{-\infty}^{\infty} x(\tau) \phi_j(T - t + \tau) \, d\tau. \qquad (5.146)$$

At time $t = T$,

$$y_j(T) = \int_{-\infty}^{\infty} x(\tau) \phi_j(\tau) \, d\tau.$$

Then,

$$y_j(T) = \int_{0}^{T} x(\tau) \phi_j(\tau) \, d\tau. \qquad (5.147)$$

$\square$

To learn the effectiveness of receiver, we shall evaluate the probability of error associated with the receiver. We continue the concept of partitioning observation space Z into M regions $\{Z_i\}_{i=1}^{M}$, following the ML decision rule. Suppose that the signal m_i is transmitted and the observation vector $\mathbf{x}$ is received from the channel. The error occurs when $\mathbf{x}$ does not fall into Z_i associated with m_i:

$$
\begin{aligned}
P_e &= \sum_{i=1}^{M} p_i P(\mathbf{x} \text{ does not lie in } Z_i | m_i \text{ sent}) \\
&= \frac{1}{M} \sum_{i=1}^{M} P(\mathbf{x} \text{ does not lie in } Z_i | m_i \text{ sent}) \\
&= 1 - \frac{1}{M} \sum_{i=1}^{M} P(\mathbf{x} \text{ lies in } Z_i | m_i \text{ sent}).
\end{aligned} \tag{5.148}
$$

We can re-write the equation in terms of likelihood function as the following (likewise N-dimensional) integral equation:

$$
P_e = 1 - \frac{1}{M} \sum_{i=1}^{M} \int_{Z_i} f_X(\mathbf{x}|m_i) \, d\mathbf{x}. \tag{5.149}
$$

As a summary, via partitioning observation space into regions $Z_1, Z_2, \ldots, Z_M$, the ML detection of a signal in AWGN is uniquely defined by the signal constellation, which has the error rate derived as above. We may conclude as follows:

Proposition 5.11. In ML detection of signal in AWGN, the probability of (symbol) error solely depends on the Euclidean distance between the message signal constellation points.

Lemma 5.12. AWGN is spherically symmetric in signal space. The rotated noise vector is also Gaussian with zero mean. Furthermore, the components of rotated noise vector are uncorrelated and thus independent.

Proof. Let $\mathbf{I}$ denote an identity matrix. The N-dimensional signal vector $\mathbf{s}_i$ multiplying an $N \times N$ orthonormal (rotating) matrix $\mathbf{Q}$ means an rotation operation, where

$$\mathbf{Q}\mathbf{Q}^T = \mathbf{I}. \tag{5.150}$$

The rotated signal vector is

$$\mathbf{s}_{i,\text{ rotate}} = \mathbf{Q}\mathbf{s}_i, \quad i = 1, 2, \ldots, M. \tag{5.151}$$

The corresponding noise vector $\mathbf{w}$ after rotation is given by

$$\mathbf{w}_{\text{rotate}} = \mathbf{Q}\mathbf{w}. \tag{5.152}$$

Since linear transformation (or combination) of Gaussian random variables is jointly Gaussian, $\mathbf{w}_{\text{rotate}}$ is Gaussian with mean

$$\begin{aligned} E[\mathbf{w}_{\text{rotate}}] &= E[\mathbf{Q}\mathbf{w}] \\ &= \mathbf{Q}E[\mathbf{w}] \\ &= \mathbf{0}. \end{aligned} \tag{5.153}$$

The covariance matrix of the noise vector is given by

$$E[\mathbf{w}\mathbf{w}^\mathrm{T}] = \frac{N_0}{2}\mathbf{I}. \tag{5.154}$$

Consequently, the covariance matrix of the rotated noise vector $\mathbf{w}_{\text{rotate}}$ is given by

$$\begin{aligned} E[\mathbf{w}_{\text{rotate}}\mathbf{w}_{\text{rotate}}^\mathrm{T}] &= E[\mathbf{Q}\mathbf{w}(\mathbf{Q}\mathbf{w})^\mathrm{T}] \\ &= E[\mathbf{Q}\mathbf{w}\mathbf{w}^\mathrm{T}\mathbf{Q}^\mathrm{T}] \\ &= \mathbf{Q}E[\mathbf{w}\mathbf{w}^\mathrm{T}]\mathbf{Q}^\mathrm{T} \\ &= \frac{N_0}{2}\mathbf{Q}\mathbf{Q}^\mathrm{T} \\ &= \frac{N_0}{2}\mathbf{I}. \end{aligned} \tag{5.155}$$

$\square$

The observation vector for the rotated signal constellation can be expressed as

$$\mathbf{x}_{\text{rotate}} = \mathbf{Q}\mathbf{s}_i + \mathbf{w}, \quad i = 1, 2, \ldots, M. \tag{5.156}$$

As the decision rule of ML detection depend solely on the Euclidean distance between $\mathbf{x}_{\text{rotate}}$ and $\mathbf{s}_{i,\text{ rotate}}$, it is straightforward to conclude that

$$\|\mathbf{x}_{\text{rotate}} - \mathbf{s}_{i,\text{ rotate}}\| = \|\mathbf{x} - \mathbf{s}_i\| \quad \text{for all } i. \tag{5.157}$$

Proposition 5.13 (Principle of Rotational Invariance). If a signal constellation is rotated by an orthonormal transformation $\mathbf{s}_{i,\text{ rotate}} = \mathbf{Q}\mathbf{s}_i$, $i = 1, 2, \ldots, M$, the probability of error associated with ML detection in AWGN remains the same.

Proposition 5.14 (Principle of Translation Invariance). If a signal constellation is translated by a constant vector, the probability of error associated with ML detection in AWGN remains the same.

Proof. Suppose that all message points in a signal constellation are translated by a constant vector $\mathbf{c}$. That is,

$$\mathbf{s}_{i,\text{ translate}} = \mathbf{s}_i - \mathbf{c}, \quad i = 1, 2, \ldots, M. \tag{5.158}$$

The observation vector is correspondingly translated by the same constant vector as

$$\mathbf{x}_{\text{translate}} = \mathbf{x} - \mathbf{c}. \tag{5.159}$$

It is obvious to reach

$$\|\mathbf{x}_{\text{translate}} - \mathbf{s}_{i,\text{ translate}}\| = \|\mathbf{x} - \mathbf{s}_i\| \quad \text{for all } i. \tag{5.160}$$

$\square$

Example 5.4. Figure 5.25 illustrates a case of rotational invariance. Figure 5.26 depicts another case of translational invariance.

As mentioned in Chapter 1, many challenges in communication system design lie in power-limited channel, and thus the design of

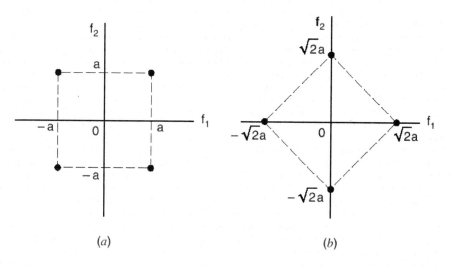

Fig. 5.25 Rotational invariance of signal constellations.

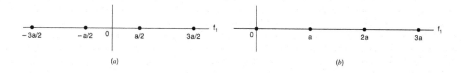

Fig. 5.26 Translational invariance of signal constellation.

the minimum energy signals under the same probability of error (using Propositions 5.13 and 5.14) is very often an important task. To inspect this issue, we again consider a set of signals $m_1, m_2, \ldots, m_M$ represented by the signal vectors $s_1, s_2, \ldots, s_M$. The average energy of this signal constellation translated by a vector $\mathbf{a}$ with p_i as *a priori* probability for m_i is

$$E_{\text{translate}} = \sum_{i=1}^{M} \|s_i - \mathbf{a}\|^2 p_i. \tag{5.161}$$

The squared Euclidean distance between s_i and $\mathbf{a}$ is given by

$$\|s_i - \mathbf{a}\|^2 = \|s_i\|^2 - 2\mathbf{a}^T s_i + \|\mathbf{a}\|^2.$$

Therefore, the average energy is expressed as

$$\mathsf{E}_{\text{translate}} = \sum_{i=1}^{M} \|\mathbf{s}_i\|^2 p_i - 2 \sum_{i=1}^{M} \mathbf{a}^{\mathrm{T}} \mathbf{s}_i p_i + \|\mathbf{a}\|^2 \sum_{i=1}^{M} p_i$$

$$= \mathsf{E} - 2\mathbf{a}^{\mathrm{T}} E[\mathbf{s}] + \|\mathbf{a}\|^2, \tag{5.162}$$

where $\mathsf{E} = \sum_{i=1}^{M} \|\mathbf{s}_i\|^2 p_i$ is the average energy of the original signal constellation, and

$$E[\mathbf{s}] = \sum_{i=1}^{M} \mathbf{s}_i p_i. \tag{5.163}$$

The minimization of (5.162) gives the translation to yield minimum energy as

$$\mathbf{a}_{\min} = E[\mathbf{s}]. \tag{5.164}$$

The resulting minimum energy is given by

$$\mathsf{E}_{\text{translate,min}} = \mathsf{E} - \|\mathbf{a}_{\min}\|^2. \tag{5.165}$$

Proposition 5.15 (Minimum Energy Translation). Given a signal constellation $\{s_i(t)\}_{i=1}^{M}$, the signal constellation achieves the minimum average energy by translating $\mathbf{s}_i$ to a constant $E[\mathbf{s}]$ as (5.164). If we further consider *a priori* probability $\{p_i\}_{i=1}^{M}$, it is equivalent to translation to the *center of (probability) mass*.

It is difficult to calculate the exact expressions of probability of error in many cases. However, we may still wish to obtain the knowledge regarding the probability of error. The common approaches to resolve this dilemma are either to find a good approximation or to obtain the performance bounds (upper bound and lower bound). The upper bound of probability of error is of particular interests, because it provides the worst-case scenario. In what follows, we present a useful upper bound (known as *union bound*) for a set of M equally likely signals in AWGN channel.

Let $A_{ik} i, k = 1, 2, \ldots, M$, denote the event that the observation vector $\mathbf{x}$ is closer to the signal vector $\mathbf{s}_k$ than to $\mathbf{s}_i$, given m_i is sent. The conditional probability of error given m_i, $P_e(m_i)$, is equal to the probability of the union of events $A_{i1}, A_{i2}, \ldots, A_{iM}$ except A_{ii}. We may express

$$P_e(m_i) \leq \sum_{\substack{k=1 \\ k \neq i}}^{M} P(A_{ik}), \quad i = 1, 2, \ldots, M. \tag{5.166}$$

By introducing the pair-wise error probability $P_2(\mathbf{s}_i, \mathbf{s}_k)$,

$$P_e(m_i) \leq \sum_{\substack{k=1 \\ k \neq i}}^{M} P_2(\mathbf{s}_i, \mathbf{s}_k), \quad i = 1, 2, \ldots, M. \tag{5.167}$$

Recall that the probability of error solely depends on the Euclidean distance. By defining Euclidean distance between $\mathbf{s}_i$ and $\mathbf{s}_k$,

$$d_{ik} = \|\mathbf{s}_i - \mathbf{s}_k\|. \tag{5.168}$$

Then,

$$P_2(\mathbf{s}_i, \mathbf{s}_k) = P(\mathbf{x} \text{ is closer to } \mathbf{s}_k \text{ than } \mathbf{s}_i, \text{ when } \mathbf{s}_i \text{ is sent})$$

$$= \int_{d_{ik}/2}^{\infty} \frac{1}{\sqrt{\pi N_0}} \exp\left(-\frac{v^2}{N_0}\right) dv. \tag{5.169}$$

That is,

$$P_2(\mathbf{s}_i, \mathbf{s}_k) = Q\left(\frac{d_{ik}/2}{\sqrt{N_0/2}}\right). \tag{5.170}$$

We therefore have

$$P_e(m_i) \leq \sum_{\substack{k=1 \\ k \neq i}}^{M} Q\left(\frac{d_{ik}/2}{\sqrt{N_0/2}}\right), \quad i = 1, 2, \ldots, M. \tag{5.171}$$

The average probability of error (over M symbols with *a priori* probability p_i) is upper bounded by

$$P_e = \sum_{i=1}^{M} p_i P_e(m_i)$$

$$\leq \sum_{i=1}^{M} \sum_{\substack{k=1 \\ k \neq i}}^{M} p_i \, Q\left(\frac{d_{ik}/2}{\sqrt{N_0/2}}\right). \tag{5.172}$$

Two special cases are of particular interests in this general derivation of probability of error upper bound for general ML detection in AWGN:

(a) Suppose the signal constellation is circularly symmetric about the origin and equally probable (i.e., $p_i = 1/M$),

$$P_e \leq \sum_{\substack{k=1 \\ k \neq i}}^{M} Q\left(\frac{d_{ik}/2}{\sqrt{N_0/2}}\right) \quad \forall i, k. \tag{5.173}$$

(b) Define

$$d_{\min} = \min_{k \neq i} d_{ik} \quad \forall i, k. \tag{5.174}$$

Then,

$$P_e \leq (M-1)Q\left(\frac{d_{\min}/2}{\sqrt{N_0/2}}\right). \tag{5.175}$$

Also note that the above general derivations are for the probability of (symbol) error that is different from the probability of bit error. By assuming that all symbol errors are equally likely,

$$\text{BER} = \frac{M/2}{M-1}P_e. \tag{5.176}$$

5.6 Commonly Applied Signaling

Amplitude shift keying (ASK), frequency shift keying (FSK), and phase shift keying (PSK) are three fundamental signaling or modulation method, corresponding to information bearing into amplitude, frequency, and phase. The performance evaluations are given in Exercises 1–3. It suggests that performance is directly related to separation (more precisely, distance) between signal constellation points.

Note that QPSK (Figure 5.27) is the superposition of I-channel and Q-channel transmission, which are orthogonal. It is equivalent to two parallel binary PSK (BPSK) transmissions in the channel. It suggests that $p_{e,\text{QPSK}} \equiv p_{e,\text{BPSK}}$.

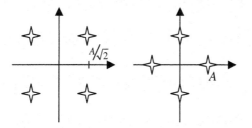

Fig. 5.27 QPSK constellation.

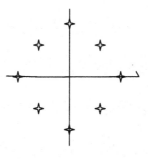

Fig. 5.28 8-PSK constellation.

We may have even higher-dimension signal constellation for PSK, and Figure 5.28 illustrates 8-PSK. There exists only $\pi/4$ phase difference among 8-PSK signal constellations, which suggests neither antipodal nor orthogonal. We can find the error rate bound in Chapter 6 to learn much poorer performance compared with BPSK and binary FSK (BFSK), however, with much more spectral efficiency (i.e., 3 bits per symbol).

To reach even higher spectral efficiency, Figure 5.29 illustrates a good example known as 16-APSK (amplitude-PSK) that is used in DVB-S2. 16-APSK delivers 4 bits per symbol spectral efficiency, even higher than 8-PSK and at the price of even poorer performance.

Although PSK has good performance, it requires precise phase recovery in demodulation, which is known as *coherent communication* and is difficult when transmission speed is high (higher bandwidth is easier to result in severe signal fading). We may use FSK instead as

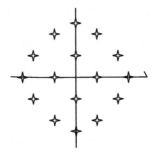

Fig. 5.29 16-APSK constellation.

FSK has two kinds of demodulations:

- Coherent demodulation to require perfect recovery of frequency and phase.
- Non-coherent demodulation to determine the maximum energy between two frequency sub-bands.

Such a flexibility makes FSK modulation attractive at one time. However, with advance of microelectronics, we can afford more and more complexity to make coherent demodulation more common in recent years. Multi-level FSK is also possible as shown in Figure 5.30, by frequency domain PAM. To yield better performance, we may want to keep the signal bands in frequency-domain PAM non-overlapping (i.e., orthogonal signaling) as shown in Figure 5.30. Sometimes, we may want to take spectral efficiency into consideration, likely due to regulations, such that these signal bands may be overlapping to lose orthogonality. A good example is Gaussian-filtered FSK in Bluetooth.

Before explore even higher-dimensional signaling, let us look at the average probability of error (symbol) to be determined by

$$p_e = \sum_{i=1}^{M} p_i P(\aleph \notin z_i | m_i) = \frac{1}{M} \sum_{i=1}^{M} P(\aleph \notin z_i | m_i) = 1 - \frac{1}{M} \sum_{i=1}^{M} P(\aleph \in z_i | m_i).$$

For 4-level PAM with equally probable assumption, we can calculate as 4-level ASK. Note here that the left-most and right-most signal constellations can have errors only in one direction (i.e., one decision

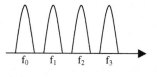

Fig. 5.30 Frequency-domain PAM.

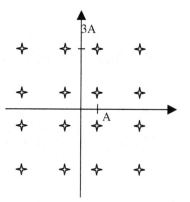

Fig. 5.31 16-QAM.

boundary, instead of two as usual). With roughly the same error rate in terms of $\frac{E_b}{N_0}$, 4-level PAM uses much power for transmission than QPSK.

We most widely used high-spectral efficient modulation is quadrature-amplitude modulation (QAM) that requires coherent demodulation. Figure 5.31 illustrates 16-QAM that carries 4 bits per symbol. The error rate analysis of 16-QAM follows the procedure:

(a) To calculate the average energy per symbol, which is $E_b = 5A^2$;

(b) I- and Q-channels introduce iid noise;

(c) Symbol error rate is therefore approximately $p_e = \frac{3}{2}Q\left(\sqrt{\frac{E_b/5}{N_0/2}}\right)$.

For even higher-dimensional QAM such as 64-QAM and 256-QAM, we follow the same principle. QAM is our common selection for modulation toward high bandwidth efficiency.

Up to this point, we may note that the performance of modulation/signaling in AWGN is pretty much related to its geographical

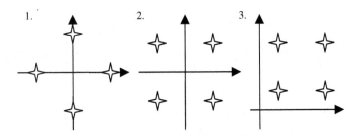

Fig. 5.32 Case1, Case 2, Case 3 of equal error rate.

distance. Consequently, in three cases of Figure 5.32, in spite of different energies per symbol, the error rate should be the same but case 3 has higher energy.

In order to minimize energy (to save transmission power), we shall have the condition of

$$\sum_i \rho_i s_i = 0. \tag{5.177}$$

It is just like the center of gravity in physics, if we treat the *a priori* probability ρ_i as mass. Minimizing $||r - s||^2$ suggests that p_e is *rotationally invariant, translation invariant*, since p_e depends only on Euclidean distance in AWGN.

Finally, for general signaling, we have to transform time-domain/frequency-domain waveforms into signal space by using the fact that therefore $\int_{-\infty}^{\infty} \phi_i(t)\phi_j(t)dt = \delta_{ij}$ implies orthogonal waveform. For example, we can transform the following waveform into a new coordinate of signal space to evaluate the performance, which is common to design signaling in communications.

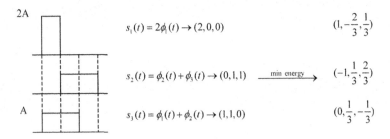

Exercises

1. The following signaling represents OOK (on–off keying) that is a special case of ASK, and find the BER with embedded AWGN $n(t)$ with zero mean and variance N_0

$$\begin{cases} \text{``1''} & A\sqrt{2}\cos\omega t \\ \text{``0''} & 0 \end{cases} + n(t).$$

2. The following signaling represents BFSK with embedded AWGN $n(t)$ with zero mean and variance N_0 and large enough frequency separation (i.e., $|\omega_1 - \omega_2|$) to maintain signal orthogonality. In other words, BFSK is a sort of orthogonal signaling.

$$\begin{cases} \text{``1''} & A\sqrt{2}\cos\omega_1 t \\ \text{``0''} & A\sqrt{2}\cos\omega_2 t \end{cases} + n(t).$$

   (a) Find the signal space and compare with ASK.

   (b) Find the BER and compare with ASK.

3. The next signaling is PSK as follows, with embedded AWGN $n(t)$ with zero mean and variance N_0. Draw the signal space to illustrate that BPSK is a sort of antipodal signaling and find the BER. What kind of conclusion can you make by Exercises 1–3? Hint: consider the signal separation

$$\begin{cases} \text{``1''} & A\sqrt{2}\cos\omega t \\ \text{``0''} & A\sqrt{2}\cos(\omega t + \tau) \end{cases} + n(t).$$

4. Consider the two 8-point QAM signal constellation shown in the following figure. The minimum distance between adjacent points is 2A. Determine the average transmitted power for each constellation assuming that the signal points are equally probable. Which constellation is more power efficient?

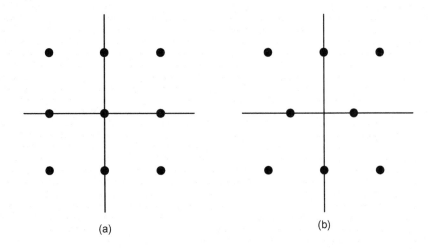

(a)   (b)

5. Consider a digital communication system that transmits information via QAM over a voice-band telephone channel at a rate of
   2400 symbols/second. The additive noise is assumed to be white
   and Gaussian

   (a) Determine the ε_b/N_0 required to achieve an error
       probability of 10^{-5} at 4800 bps.

   (b) Repeat (a) for a bit rate of 9600 bps.

   (c) Repeat (a) for a bit rate of 19200 bps.

   (d) What conclusions do you reach from these results.

6. Consider the 8-point QAM signal constellation shown in the
   Figure

   (a) Is it possible to assign three data bits to each point of
       the signal constellation such that the nearest (adja-
       cent) points differ in only one bit position?

   (b) Determine the symbol rate if the desired bit rate is
       90 Mbps.

   (c) Compare the *SNR* required for the 8-point QAM
       modulation with that of an 8-point PSK modulation
       having the same error probability.

(d) Which signal constellation, 8-point QAM or 8-point PSK, is more immune to phase errors? Explain the reason for your answer.

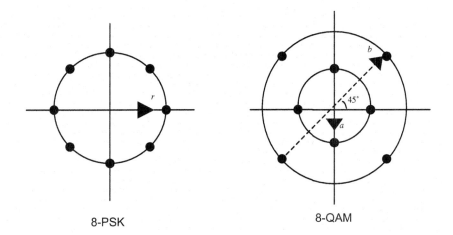

8-PSK   8-QAM

7. The following waveforms of signal $s_i(t)$, $i = 1, 2, 3, 4$

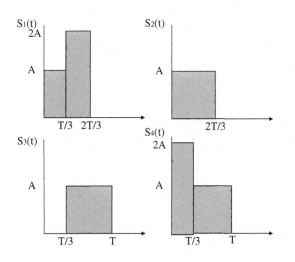

(a) Using Gram–Schmidt procedure to find an orthonormal basis for the set of signals.

(b) Construct the corresponding signal–space diagram.

(c) Find the minimum energy signaling in this case and thus the corresponding signal–space diagram.

(d) If there is an embedded AWGN noise with zero mean and two-side psd $N_0/2$, find the optimal receiver and the associated BER in (c).

8. This is regarding 2D matched filter

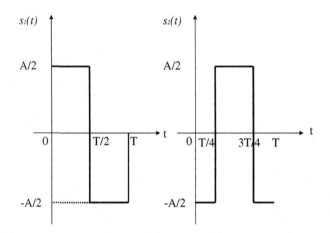

(a) Determine the impulse response of matched filter for each of these two signals.

(b) Also plot the matched filter output as a function of time for these two signals.

(c) Are these two signals orthogonal to each other?

(d) By connecting two matched filters in parallel (just like parallel as electronic circuits), can we design a detector to distinguish these two signals?

(e) In (d), if $s_1(t)$ and $s_2(t)$ are transmitted equally probably in an AWGN with zero mean and psd $N_0/2$, derive the error probability.

(f) Is this detector in (e) optimal? If yes, why? If not, design the optimal detector.

9. The following waveforms of signal $s_i(t)$, $i = 1, 2, 3, 4$.

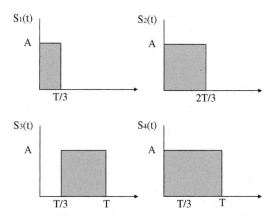

(a) Using Gram–Schmidt procedure find an orthonormal basis for the set of signals.

(b) Construct the corresponding signal–space diagram.

(c) Find the minimum energy signaling in this case, and thus the corresponding signal–space diagram.

(d) If there is an embedded AWGN noise with zero mean and two-side psd $N_0/2$, find the BER in (c).

6

Passband Digital Transmission

6.1 Digital Modulations

We have discussed baseband pulse transmission in Chapter 4. After introduction of optimal receiver principles, we are studying passband digital transmission, that is, using higher frequency carrier transmits digital data. Such digital information can be generally stored in amplitude, frequency, and phase, the same principle as baseband pulse transmission. Figure 6.1 illustrates the way to embed digital data (i.e., "1" and "0") into waveforms. From here, we start from digital modulation suitable for passband digital transmission, which can be categorized into three classes:

- Amplitude-shift keying (ASK);
- Frequency-shift keying (FSK);
- Phase-shift keying (PSK).

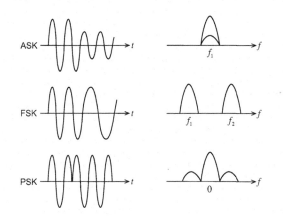

Fig. 6.1 MSK phase change.

195

The above digital modulations can be in a general *M-ary* signaling. For ASK, it becomes *M*-ary pulse amplitude modulation (PAM), and we can implement PAM into both in-phase (I) channel and quadrature-phase (Q) channel to form *quadrature-amplitude modulation* (QAM). Also note that both PSK and FSK have the desirable *constant-envelope* property. Digital modulations can be categorized into *coherent* and *noncoherent* classes. In case locally generated oscillator reference is employed to fully synchronize received waveforms (or to exactly recover phase), such modulations are *coherent* modulations. PSK-type modulations are surely coherent, except *differential PSK* (DPSK), as a non-coherent modulation need not exactly recover phase but only need to recover phase-difference.

To completely understand digital modulation, its frequency domain features are important, especially its *power spectra*. Figure 6.1 also illustrates the power spectra of ASK, FSK, and PSK. Power spectral density (psd) of digital modulations originates from its randomness of random data.

For example, the psd of binary PSK (BPSK) can be calculated using the following steps:

(a) Random phase from the random process generated by random data,

$$\Phi(t) = \begin{cases} 0 \text{ with prob. } \frac{1}{2} \\ \pi \text{ with prob. } \frac{1}{2}; \end{cases} \tag{6.1}$$

(b) $s(t) = \sqrt{2}\cos(2\pi f_c t + \Phi(t))$;
(c) $R_s(\tau) = E[s(t + \tau)s(t)]$, assume wide-sense stationary;
(d) $S_s(f) = \mathcal{F}\{R_s(\tau)\}$, which consists of $\text{sinc}^2(f)$ from $\Phi(t)$.

The bandwidth defined by the first-null of psd represents the signal bandwidth of BPSK modulation. Suppose the data rate is R_b and the effectively used channel bandwidth is B, the bandwidth efficiency

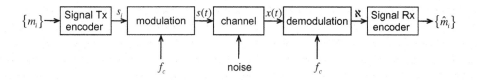

Fig. 6.2 Passband transmission model.

(or signal spectral efficiency), ρ, is given by

$$\rho = \frac{R_b}{B} \text{ bits/second/Hz.} \tag{6.2}$$

The passband transmission system is depicted in Figure 6.2. The message source emits one symbol every T seconds, with the symbol for belonging to an alphabet of size M (i.e., $m_1, m_2, \ldots, m_M$ with *a priori* probabilities $P(m_1), \ldots, P(m_M)$). For equally probable signaling, $p(m_i) = \frac{1}{M}$.

The signal energy of m_i is $E_i = \int_{-\infty}^{\infty} s_i^2(t)dt$. The channel is assumed linear and has enough bandwidth to avoid signal distortion. The additive channel noise $w(t)$ is white and Gaussian with zero mean and psd $N_0/2$. The receiver consists of a detector followed by a signal transmission decoder to estimate $\hat{m}$ by minimizing noise effect.

6.1.1 Coherent PSK

For a coherent BPSK,

$$s_1(t) = \sqrt{\frac{2E_b}{T_b}} \cos(2\pi f_c t), \tag{6.3}$$

$$s_0(t) = \sqrt{\frac{2E_b}{T_b}} \cos(2\pi f_c t + \pi) = -\sqrt{\frac{2E_b}{T_b}} \cos(2\pi f_c t), \tag{6.4}$$

where $0 \leq t \leq T_b$ and E_b is the transmitted signal energy per bit. Such a signaling is antipodal. Following signal space methodology,

$$\phi_1(t) = \sqrt{\frac{2}{T_b}} \cos(2\pi f_c t), \quad 0 \leq t \leq T_b. \tag{6.5}$$

We therefore have

$$s_1(t) = \sqrt{E_b}\phi_1(t), \qquad 0 \le t \le T_b, \tag{6.6}$$
$$s_0(t) = -\sqrt{E_b}\phi_1(t), \qquad 0 \le t \le T_b. \tag{6.7}$$

The message points in signal space (1-dimension) have coordinates

$$s_{11} = \int_0^{T_b} s_1(t)\phi_1(t)dt = +\sqrt{E_b}, \tag{6.8}$$

$$s_{01} = \int_0^{T_b} s_0(t)\phi_1(t)dt = -\sqrt{E_b}. \tag{6.9}$$

Figure 6.3 illustrates the signal space diagram.

The maximum likelihood (ML) decision is based on the received signal point falling into Z_1 or Z_0, while the decision boundary between Z_1 and Z_0 is set to be 0. The conditional probability of error deciding

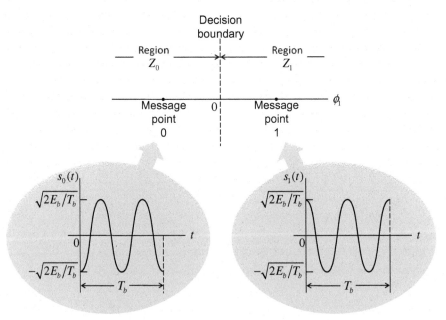

Fig. 6.3 BPSK signal space.

"1" given transmitting "0" is given by

$$p_{10} = \int_0^\infty f_{X_1}(x_1|0)dx_1$$

$$= \int_0^\infty \frac{1}{\sqrt{\pi N_0}} e^{-\frac{1}{N_0}(x_1+\sqrt{E_b})^2} dx_1$$

$$= Q\left(\sqrt{\frac{2E_b}{N_0}}\right). \tag{6.10}$$

The average probability of symbol error (or bit error rate, BER) is thus given by

$$P_e = Q\left(\sqrt{\frac{2E_b}{N_0}}\right). \tag{6.11}$$

Consequently, Figure 6.4 shows the block diagram of a BPSK transmitter and receiver. We may note the receiver structure as a correlation receiver or a matched filter.

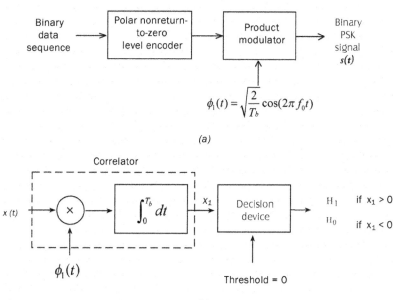

Fig. 6.4 BPSK: (a) transmitter; (b) receiver.

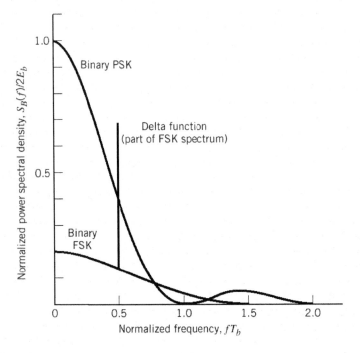

Fig. 6.5 Power spectra of BPSK and BFSK signals.

The power spectrum of BPSK is as follows and is plotted in Figure 6.5:

$$S_B(f) = 2E_b \sin c^2(T_b f). \tag{6.12}$$

Quadrature PSK (QPSK) develops in a similar manner as BPSK. Figure 6.6 shows the block diagrams of QPSK transmitter and receiver. Figure 6.7 depicts the power spectrum of QPSK.

QPSK has the optimum BER performance and twice the spectral efficiency than BPSK. However, there is a major drawback as shown in Figure 6.8, especially for power-limited channels, phase discontinuity. The solution gives a new modulation called *offset QPSK* (OQPSK), by offsetting I-channel data transmission and Q-channel data transmission by half symbol period/duration. Therefore, the phase can change at most 90° by avoiding simultaneous bit transitions at I and Q channels that results in 180° phase change.

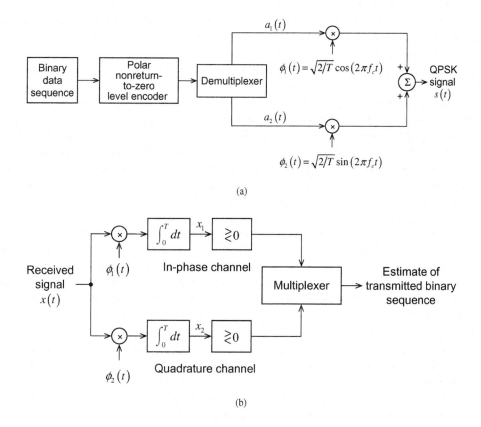

Fig. 6.6 QPSK: (a) transmitter; (b) receiver.

Another variation of QPSK to achieve similar purpose of OQPSK is $\pi/4$-QPSK by picking up one of the two QPSK constellations of Figure 6.9 to yield eight possible phase states as shown in Figure 6.10.

$\pi/4$-QPSK has an obvious attractive feature, phase transition restricted to $\pm\pi/4$ and $\pm3\pi/4$, which results in significant reduction in envelope variation. However, $\pi/4$-QPSK is in fact differentially coded and thus noncoherent detection is possible to suggest less complexity of receiver. The detection of $\pi/4$-QPSK signal is shown in Figure 6.11.

For M-ary $(M\geq4)$ PSK signaling, the signal space can be found as shown in Figure 6.12, and the average probability of error through the

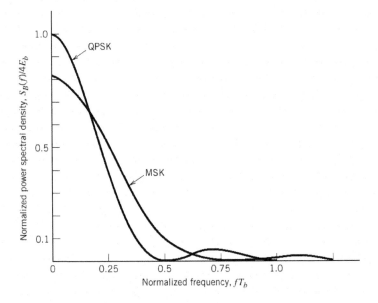

Fig. 6.7 Power spectrum of QPSK.

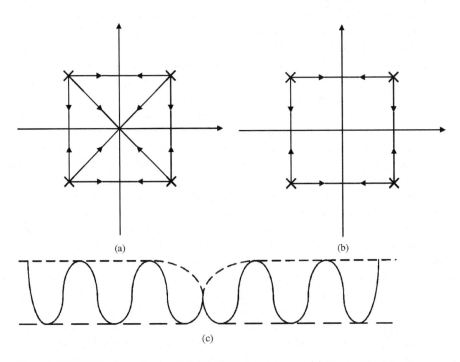

Fig. 6.8 (a) QPSK phase change; (b) OQPSK phase change; (c) phase discontinuity.

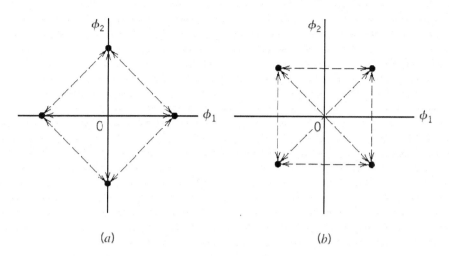

Fig. 6.9 Two constellations of QPSK.

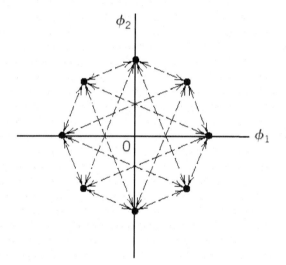

Fig. 6.10 Eight possible phase states for $\pi/4$-QPSK.

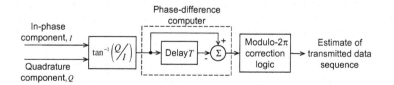

Fig. 6.11 Block diagram of a $\pi/4$-QPSK detector.

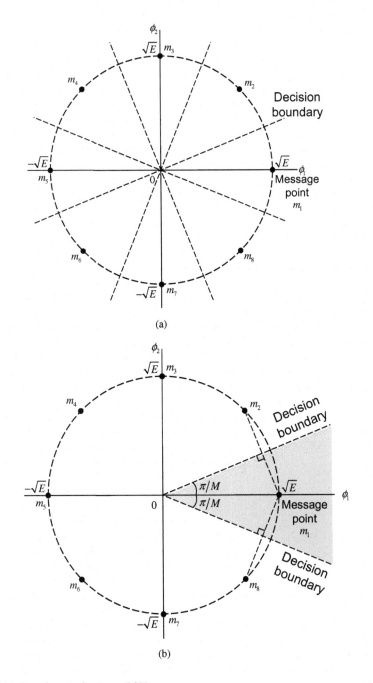

(a)

(b)

Fig. 6.12 Signal space for 8-ary PSK.

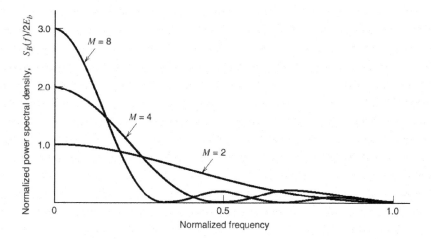

Fig. 6.13 Power spectrum of M-ary PSK.

concept of union bound is given by

$$P_e \cong 2Q\left(\sqrt{\frac{2E}{N_0}}\sin\left(\frac{\pi}{M}\right)\right). \tag{6.13}$$

The power spectrum of M-ary PSK is shown in Figure 6.13.

6.1.2   Coherent Frequency Shift Keying

Binary FSK (BFSK) system typically transmits the following wave-forms:

$$s_i(t) = \begin{cases} \sqrt{\dfrac{2E_b}{T_b}}\cos(2\pi f_i t), & 0 \le t \le T_b \\ 0, & \text{elsewhere,} \end{cases} \tag{6.14}$$

where $i = 1, 2$, and the transmitted frequency is given by

$$f_i = \frac{n_c + i}{T_b} \quad \text{for some fixed integer } n_c \quad \text{and} \quad i = 1, 2. \tag{6.15}$$

The above FSK signal is a *continuous-phase* FSK (CP-FSK) to avoid unnecessary side-lobes for signals. We shall select the following orthonormal basis functions for signal space:

$$\phi_i(t) = \begin{cases} \sqrt{\dfrac{2}{T_b}} \cos(2\pi f_i t), & 0 \le t \le T_b \\ 0, & \text{elsewhere.} \end{cases} \tag{6.16}$$

The coefficients for signal points are

$$\begin{aligned} s_{ij} &= \int_0^{T_b} s_i(t)\,\phi_j(t)\,dt \\ &= \int_0^{T_b} \sqrt{\dfrac{2E_b}{T_b}} \cos(2\pi f_i t) \sqrt{\dfrac{2}{T_b}} \cos(2\pi f_j t)\,dt \\ &= \begin{cases} \sqrt{E_b} & i = j \\ 0, & i \ne j. \end{cases} \end{aligned} \tag{6.17}$$

The resulting signal space for coherent CP-FSK is illustrated in shown in Figure 6.14, while the two signal points with distance $\sqrt{2E_b}$ are given by

$$\mathbf{s}_1 = \begin{bmatrix} \sqrt{E_b} \\ 0 \end{bmatrix}, \tag{6.18}$$

$$\mathbf{s}_2 = \begin{bmatrix} 0 \\ \sqrt{E_b} \end{bmatrix}. \tag{6.19}$$

The observation vector $\mathbf{x}$ has the following two elements:

$$x_1 = \int_0^{T_b} x(t)\,\phi_1(t)\,dt, \tag{6.20}$$

$$x_2 = \int_0^{T_b} x(t)\,\phi_2(t)\,dt. \tag{6.21}$$

Define a new Gaussian random variable Y whose sample value $y = x_1 - x_2$. Then,

$$\begin{aligned} E[Y|1] &= E[X_1|1] - E[X_2|1] \\ &= +\sqrt{E_b}, \end{aligned} \tag{6.22}$$

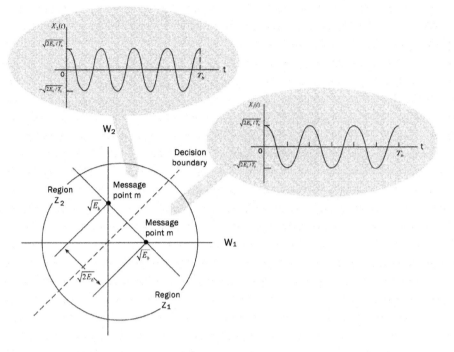

Fig. 6.14 FSK signal space.

$$E[Y|0] = E[X_1|0] - E[X_2|0]$$
$$= -\sqrt{E_b}, \tag{6.23}$$
$$\text{var}[Y] = \text{var}[X_1] + \text{var}[X_2]$$
$$= N_0/2 + N_0/2 = N_0. \tag{6.24}$$

The conditional probability is expressed as

$$f_Y(y|0) = \frac{1}{\sqrt{2\pi N_0}} \exp\left[-\frac{(y + \sqrt{E_b})^2}{2N_0}\right]. \tag{6.25}$$

The conditional probability of error is given by

$$p_{10} = P(y > 0 | \text{symbol 0 was sent})$$
$$= \int_0^\infty f_Y(y|0)\, dy$$

$$= \frac{1}{\sqrt{2\pi N_0}} \int_0^\infty \exp\left[-\frac{(y + \sqrt{E_b})^2}{2N_0}\right] dy$$

$$= \frac{1}{\sqrt{2\pi}} \int_{\sqrt{E_b/N_0}}^\infty \exp\left[-\frac{z^2}{2}\right] dz$$

$$= Q\left(\sqrt{\frac{E_b}{N_0}}\right). \tag{6.26}$$

Assuming equally probably BFSK, the average probability of error is given by

$$P_e = Q\left(\sqrt{\frac{E_b}{N_0}}\right). \tag{6.27}$$

Comparing (6.27) with (6.11), we observe 3 dB performance gain for BPSK over BFSK. It suggests that antipodal signal is superior to orthogonal signaling by 3 dB.

Figure 6.15 depicts the FSK transceiver (transmitter and receiver). To maintain the phase continuity, BFSK signal can be expressed as

$$s(t) = \sqrt{\frac{2E_b}{T_b}} \cos\left(2\pi f_c t \pm \frac{\pi t}{T_b}\right), \quad 0 \le t \le T_b. \tag{6.28}$$

By trigonometric identity, we have

$$s(t) = \sqrt{\frac{2E_b}{T_b}} \cos\left(\pm\frac{\pi t}{T_b}\right) \cos\left(2\pi f_c t\right) - \sqrt{\frac{2E_b}{T_b}} \sin\left(\pm\frac{\pi t}{T_b}\right) \sin\left(2\pi f_c t\right)$$

$$= \sqrt{\frac{2E_b}{T_b}} \cos\left(\frac{\pi t}{T_b}\right) \cos\left(2\pi f_c t\right) \mp \sqrt{\frac{2E_b}{T_b}} \sin\left(\frac{\pi t}{T_b}\right) \sin\left(2\pi f_c t\right). \tag{6.29}$$

In (6.29), the in-phase component is completely independent of the input binary waveform. Its power spectral density consists of two spikes (delta functions) at $f = \pm 1/2T_b$. The quadrature component is directly related to the input binary waveform. Define the symbol shape function as

$$g(t) = \begin{cases} \sqrt{\frac{2E_b}{T_b}} \sin\left(\frac{\pi t}{T_b}\right), & 0 \le t \le T_b \\ 0, & \text{elsewhere.} \end{cases} \tag{6.30}$$

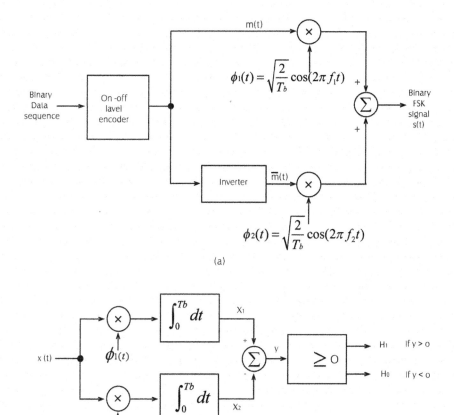

(a)

(b)

Fig. 6.15 (a) FSK transmitter; (b) FSK receiver.

Its energy spectral density is then given by

$$\Psi_g(f) = \frac{8E_b T_b \cos^2(\pi T_b f)}{\pi^2 (4T_b^2 f^2 - 1)^2}.$$ (6.31)

Consequently, the overall psd is expressed as

$$S_B(f) = \frac{E_b}{2T_b}\left[\delta\left(f - \frac{1}{2T_b}\right) + \delta\left(f + \frac{1}{2T_b}\right)\right] + \frac{8E_b \cos^2(\pi T_b f)}{\pi^2 (4T_b^2 f^2 - 1)^2}.$$ (6.32)

In what follows, we are going to exploit performance improvement by properly using phase in detection. Consider a CP-FSK signal as follows:

$$s(t) = \begin{cases} \sqrt{\frac{2E_b}{T_b}} \cos[2\pi f_1 t + \theta(0)] & \text{for symbol 1} \\ \\ \sqrt{\frac{2E_b}{T_b}} \cos[2\pi f_2 t + \theta(0)] & \text{for symbol 0.} \end{cases} \tag{6.33}$$

Another conventional form to represent the CP-FSK signal is through the angle-modulated signal concept, which is given by

$$s(t) = \sqrt{\frac{2E_b}{T_b}} \cos[2\pi f_c t + \theta(t)]. \tag{6.34}$$

The phase $\theta(t)$ of a CP-FSK signal increases or decreases linearly with time as

$$\theta(t) = \theta(0) \pm \frac{\pi h}{T_b} t, \quad 0 \le t \le T_b, \tag{6.35}$$

where "+" means transmitting "1" and "−" means transmitting "0", and h is later shown as the *deviation ratio*. We can derive the following relationship:

$$f_c + \frac{h}{2T_b} = f_1, \tag{6.36}$$

$$f_c - \frac{h}{2T_b} = f_2. \tag{6.37}$$

Define

$$f_c = \frac{1}{2}(f_1 + f_2), \tag{6.38}$$

$$h = T_b(f_1 - f_2). \tag{6.39}$$

For $t = T_b$,

$$\theta(T_b) - \theta(0) = \begin{cases} \pi h & \text{for symbol "1"} \\ -\pi h & \text{for symbol "0".} \end{cases} \tag{6.40}$$

We can form a phase trellis as shown in Figure 6.16 by considering a sequence of symbols (or bits). With $h = 1/2$, frequency deviation equals half the bit rate and is the minimum frequency spacing to allow

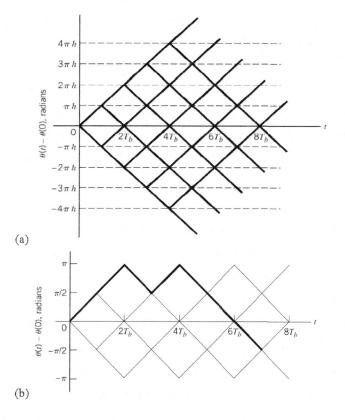

Fig. 6.16 (a) Phase trellis; (b) sequence 1101000.

no interference in the detection of two FSK signals. Such a CP-FSK is called minimum shift keying (MSK).

Note that the phase trellis structure of continuous phase modulation suggests a new way of optimal detection by decoding the entire sequence of symbols based on the phase trellis. Such a methodology is called *maximum likelihood sequence estimation* (MLSE) and can be implemented by the well-known *Viterbi algorithm* from decoding convolutional codes. For more details about MLSE see Chapter 8.

We may express the CP-FSK signal in canonical form as

$$s(t) = \sqrt{\frac{2E_b}{T_b}} \cos[\theta(t)] \cos(2\pi f_c t) - \sqrt{\frac{2E_b}{T_b}} \sin[\theta(t)] \sin(2\pi f_c t). \quad (6.41)$$

With $h = 1/2$, we have

$$\theta(t) = \theta(0) \pm \frac{\pi}{2T_b}t, \quad 0 \le t \le T_b, \tag{6.42}$$

where the plus sign represents "1" and the minus sign represents "0". Since the phase $\theta(0)$ is 0 or π depending on the past modulation process, the polarity of $\cos[\theta(t)]$ depends only on $\theta(0)$ for $t \in [-T_b, T_b]$. For $t \in [-T_b, T_b]$, the in-phase component is given by

$$
\begin{aligned}
s_I(t) &= \sqrt{\frac{2E_b}{T_b}} \cos[\theta(t)] \\
&= \sqrt{\frac{2E_b}{T_b}} \cos[\theta(0)] \cos\left(\frac{\pi}{2T_b}t\right) \\
&= \pm\sqrt{\frac{2E_b}{T_b}} \cos\left(\frac{\pi}{2T_b}t\right), \quad -T_b \le t \le T_b.
\end{aligned}
\tag{6.43}
$$

In (6.43), the plus sign corresponds to $\theta(0) = 0$ and the minus sign corresponds to $\theta(0) = \pi$. Similarly, for $t \in [0, 2T_b]$, the quadrature component is given by

$$
\begin{aligned}
s_Q(t) &= \sqrt{\frac{2E_b}{T_b}} \sin[\theta(t)] \\
&= \sqrt{\frac{2E_b}{T_b}} \sin[\theta(T_b)] \sin\left(\frac{\pi}{2T_b}t\right) \\
&= \pm\sqrt{\frac{2E_b}{T_b}} \sin\left(\frac{\pi}{2T_b}t\right), \quad 0 \le t \le 2T_b.
\end{aligned}
\tag{6.44}
$$

Here, the plus sign corresponds to $\theta(T_b) = \pi/2$ and the minus sign corresponds to $\theta(T_b) = -\pi/2$. We may use Table 6.1 to summarize the modulation mapping.

Table 6.1

$\theta(0)$	$\theta(T_b)$	Transmitted symbol
0	$\pi/2$	1
π	$\pi/2$	0
π	$-\pi/2$	1
0	$-\pi/2$	0

From (6.41), we may derive a set of orthonormal basis functions for MSK as follows:

$$\phi_1(t) = \sqrt{\frac{2}{T_b}} \cos\left(\frac{\pi}{2T_b}t\right) \cos(2\pi f_c t), \quad 0 \le t \le T_b \tag{6.45}$$

$$\phi_2(t) = \sqrt{\frac{2}{T_b}} \sin\left(\frac{\pi}{2T_b}t\right) \sin(2\pi f_c t), \quad 0 \le t \le T_b. \tag{6.46}$$

Therefore, MSK signal can be expanded as

$$s(t) = s_1\phi_1(t) + s_2\phi_2(t), \quad 0 \le t \le T_b, \tag{6.47}$$

To obtain the coefficients,

$$
\begin{aligned}
s_1 &= \int_{-T_b}^{T_b} s(t)\phi_1(t)\, dt \\
&= \sqrt{E_b}\cos[\theta(0)], \quad -T_b \le t \le T_b
\end{aligned}
\tag{6.48}
$$

$$
\begin{aligned}
s_2 &= \int_{0}^{2T_b} s(t)\phi_2(t)\, dt \\
&= -\sqrt{E_b}\sin[\theta(T_b)], \quad 0 \le t \le 2T_b.
\end{aligned}
\tag{6.49}
$$

Note from (6.48) and (6.49) that:

- Both integral intervals have a duration of $2T_b$ and have a difference of shifting (or offset) T_b in the integration range.
- The time interval $[0, T_b]$ is common to both integrals with phase states $\theta(0)$ and $\theta(T_b)$.

Consequently, the 2-D signal space diagram of MSK is shown in Figure 6.17 to present all corresponding parameters.

To evaluate the probability of error for MSK, the received waveform embedded in additive white Gaussian noise (AWGN) is given by

$$
\begin{aligned}
x_1 &= \int_{-T_b}^{T_b} x(t)\phi_1(t)\, dt \\
&= s_1 + w_1, \quad -T_b \le t \le T_b;
\end{aligned}
\tag{6.50}
$$

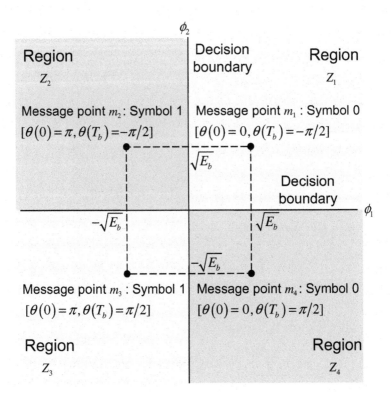

Fig. 6.17 MSK signal space.

$$x_2 = \int_0^{2T_b} x(t)\phi_2(t)\, dt$$

$$= s_2 + w_2, \quad 0 \le t \le 2T_b. \tag{6.51}$$

To reconstruct the original binary sequence, we interleave the above two sets of phase decisions as given in Figure 6.17. It is clear from signal space that the probability of error is the same as QPSK, that is

$$P_e = Q\left(\sqrt{\frac{2E_b}{N_0}}\right). \tag{6.52}$$

The generation and detection of MSK can be done in several ways. Figures 6.18 and 6.19 depict two coherent forms and it is not difficult to observe the equivalence.

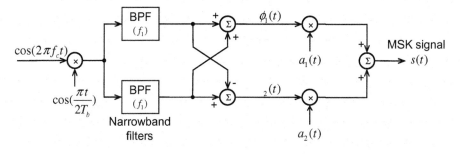

(a)

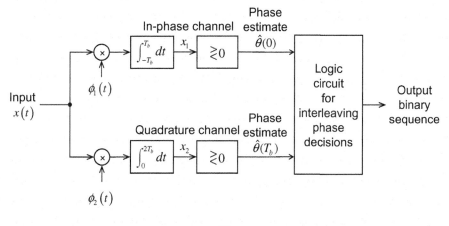

(b)

Fig. 6.18 MSK: (a) transmitter; (b) receiver.

To calculate the power spectra of MSK, depending on the phase state $\theta(0)$, the in-phase component is equal to $\pm g(t)$, that is

$$g(t) = \begin{cases} \sqrt{\dfrac{2E_b}{T_b}} \cos\left(\dfrac{\pi t}{2T_b}\right), & -T_b \leq t \leq T_b \\ 0, & \text{otherwise.} \end{cases} \tag{6.53}$$

The energy spectral density of this symbol-shaping function is expressed as

$$\psi_g(f) = \frac{32E_bT_b}{\pi^2}\left[\frac{\cos(2\pi T_bf)}{16T_b^2f^2 - 1}\right]^2. \tag{6.54}$$

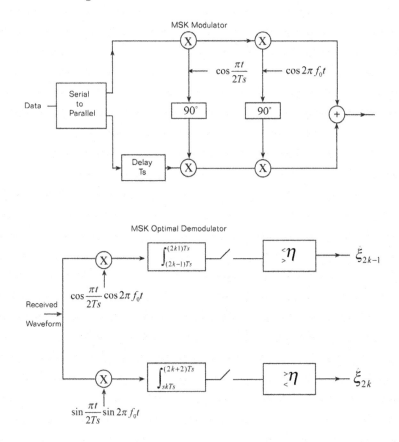

Fig. 6.19 MSK modulator and demodulator.

Depending on the phase state $\theta(0)$, the quadrature-phase component is equal to $\pm g'(t)$, which is given by

$$g'(t) = \begin{cases} \sqrt{\dfrac{2E_b}{T_b}} \sin\left(\dfrac{\pi t}{2T_b}\right), & 0 \le t \le 2T_b \\ 0, & \text{otherwise.} \end{cases} \qquad (6.55)$$

As I and Q components of MSK are statistically independent, the baseband psd of the MSK signal is expressed as

$$S_B(f) = 2\left[\frac{\psi_g(f)}{2T_b}\right]$$

$$= \frac{32E_b}{\pi^2}\left[\frac{\cos(2\pi T_b f)}{16T_b^2 f^2 - 1}\right]^2. \qquad (6.56)$$

MSK has some interesting features such as

- Constant envelope;
- Relatively good spectral efficiency;
- Coherent detection performance equivalent to QPSK;
- Non-coherent detection is possible as a sort of FSK.

To further meet the need of spectrum shape in modern wireless communications, a low-pass pulse-shaping filter is usually employed to ensure frequency response with narrow bandwidth and sharp cutoff nature and to ensure impulse response with relatively low overshoot. On the other hand, we also wish to maintain MSK phase trellis. The resulting modulation is known as *Gaussian filtered MSK* (GMSK), which is adopted in a popular second-generation digital cellular system, GSM and its evolutions (GPRS and EDGE).

Let W denote the *3 dB bandwidth* of the pulse-shaping filter. We define the transfer function $H(f)$ and impulse response $h(t)$ as follows:

$$H(f) = \exp\left(-\frac{\log 2}{2}\left(\frac{f}{W}\right)^2\right), \tag{6.57}$$

$$h(t) = \sqrt{\frac{2\pi}{\log 2}}\, W \exp\left(-\frac{2\pi^2}{\log 2}W^2 t^2\right). \tag{6.58}$$

The response of such a Gaussian filter to a rectangular pulse of unit amplitude and duration T_b is given by

$$
\begin{aligned}
g(t) &= \int_{-T_b/2}^{T_b/2} h(t-\tau)\,d\tau \\
&= \sqrt{\frac{2\pi}{\log 2}}\, W \int_{-T_b/2}^{T_b/2} \exp\left(-\frac{2\pi^2}{\log 2}W^2(t-\tau)^2\right) d\tau. \tag{6.59}
\end{aligned}
$$

Or, it may be expressed as

$$g(t) = Q\left(\frac{2\pi}{\sqrt{\log 2}}W T_b\left(\frac{t}{T_b}-\frac{1}{2}\right)\right) - Q\left(\frac{2\pi}{\sqrt{\log 2}}W T_b\left(\frac{t}{T_b}+\frac{1}{2}\right)\right). \tag{6.60}$$

As illustrated in Figure 6.20, the (time–bandwidth) product WT_b is an important design parameter for GMSK. Figure 6.21 depicts the power spectra of MSK.

The probability of error can be treated from FSK performance with a factor of α that depends on time–bandwidth product, that is,

$$P_e = Q\left(\sqrt{\frac{\alpha E_b}{N_0}}\right). \tag{6.61}$$

In what follows, we are considering the M-ary FSK, with the transmitted signals defined as

$$s_i(t) = \sqrt{\frac{2E}{T}}\cos\left[\frac{\pi}{T}(n_c + i)t\right], \quad 0 \le t \le T, \quad i = 1, 2, \ldots, M \tag{6.62}$$

Note that

$$\int_0^T s_i(t)\, s_j(t)\, dt = 0, \quad i \neq j. \tag{6.63}$$

We can therefore build up a set of orthonormal basis functions as follows:

$$\phi_i(t) = \frac{1}{\sqrt{E}} s_i(t), \quad \begin{array}{c} 0 \le t \le T \\ i = 1,\, 2,\, \ldots,\, M. \end{array} \tag{6.64}$$

The probability of error can be obtained through the union bound as

$$P_e \le (M - 1)Q\left(\sqrt{\frac{E}{N_0}}\right). \tag{6.65}$$

The power spectra of M-ary FSK is depicted in Figure 6.22. It is worth noting the channel bandwidth$B$and bandwidth efficiencyρ for M-ary FSK. To coherently detect FSK signals, the adjacent signals must separate each other by $1/2T$.

$$B = \frac{M}{2T}. \tag{6.66}$$

With $R_b = 1/T_b$, we have

$$B = \frac{R_b M}{2 \log_2 M} \tag{6.67}$$

$$\rho = \frac{R_b}{B} = \frac{2 \log_2 M}{M}. \tag{6.68}$$

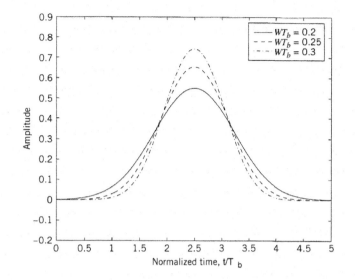

Fig. 6.20 Frequency-shaping for GMSK.

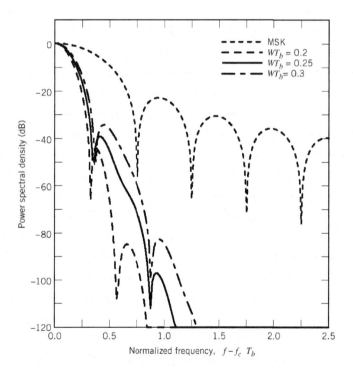

Fig. 6.21 Power spectra of MSK and GMSK.

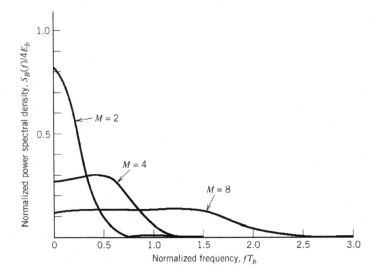

Fig. 6.22 Power spectra of M-ary FSK.

6.1.3   Non-coherent Digital Modulations

Up to this point, we assume the receiver to be perfectly synchronized to the transmitted waveform, and the only impairment is channel noise. However, it is not true in many cases, and we may have to consider noncoherent detection/demodulation of signals. A digital communication receiver that does not *exactly* recover the carrier phase is *non-coherent*.

Let us first introduce the *optimum quadratic receiver* for detection of signals with unknown phase. Consider a binary digital communication system with transmitted signal (E is the signal energy, T is the duration of signaling, f_i is an integer multiple of $1/2T$)

$$s_i(t) = \sqrt{\frac{2E}{T}} \cos(2\pi f_i t), \quad \begin{array}{c} 0 \le t \le T \\ i = 1, 2. \end{array} \tag{6.69}$$

The received non-coherent signal for an AWGN channel can be expressed as

$$x(t) = \sqrt{\frac{2E}{T}} \cos(2\pi f_i t + \theta) + w(t), \quad \begin{array}{c} 0 \le t \le T \\ i = 1, 2. \end{array} \tag{6.70}$$

It is practical and reasonable to assume no *a priori* information about the unknown carrier phase θ and to treat it as a random variable having uniform distribution

$$f_\Theta(\theta) = \begin{cases} \dfrac{1}{2\pi}, & -\pi < \theta \leq \pi \\ 0, & \text{otherwise.} \end{cases} \tag{6.71}$$

The binary detection with unknown phase (or a dummy parameter in general) becomes:

Given the received signal $x(t)$ with an unknown carrier phase θ, we want to design an optimum receiver for detecting s_i represented by the signal component $\sqrt{E/2T} \cos(2\pi f_i t + \theta)$ in $x(t)$.

The conditional likelihood function of s_i, given θ, is expressed as

$$L(s_i(\theta)) = \exp\left(\sqrt{\frac{E}{N_0 T}} \int_0^T x(t) \cos(2\pi f_i t + \theta) \, dt\right). \tag{6.72}$$

We can remove the dummy variable by averaging as

$$L(s_i) = \int_{-\pi}^{\pi} L(s_i(\theta)) \, f_\Theta(\theta) \, d\theta$$

$$= \frac{1}{2\pi} \int_{-\pi}^{\pi} \exp\left(\sqrt{\frac{E}{N_0 T}} \int_0^T x(t) \cos(2\pi f_i t + \theta) \, dt\right) d\theta. \tag{6.73}$$

Again by trigonometric formula, we obtain

$$\int_0^T x(t) \cos(2\pi f_i t + \theta) \, dt$$

$$= \cos\theta \int_0^T x(t) \cos(2\pi f_i t) \, dt - \sin\theta \int_0^T x(t) \sin(2\pi f_i t) \, dt. \tag{6.74}$$

Define

$$l_i = \left[\left(\int_0^T x(t) \cos(2\pi f_i t) \, dt\right)^2 + \left(\int_0^T x(t) \sin(2\pi f_i t) \, dt\right)^2\right]^{1/2}, \tag{6.75}$$

$$\beta_i = \tan^{-1}\left(\frac{\int_0^T x(t)\,\sin(2\pi f_i t)\,dt}{\int_0^T x(t)\,\cos(2\pi f_i t)\,dt}\right). \tag{6.76}$$

(6.74) thus becomes

$$\int_0^T x(t)\,\cos(2\pi f_i t + \theta)\,dt = l_i(\cos\theta\cos\beta_i - \sin\theta\sin\beta_i)$$

$$= l_i\cos(\theta + \beta_i) \tag{6.77}$$

Using the above equation in the average likelihood function, we have

$$L(s_i) = \frac{1}{2\pi}\int_{-\pi}^{\pi}\exp\left(\sqrt{\frac{E}{N_0 T}}\,l_i\cos(\theta + \beta_i)\right)d\theta$$

$$= \frac{1}{2\pi}\int_{-\pi+\beta_i}^{\pi+\beta_i}\exp\left(\sqrt{\frac{E}{N_0 T}}\,l_i\cos\theta\right)d\theta$$

$$= \frac{1}{2\pi}\int_{-\pi}^{\pi}\exp\left(\sqrt{\frac{E}{N_0 T}}\,l_i\cos\theta\right)d\theta. \tag{6.78}$$

Recall the *modified Bessel function of the first land of zero order*, that is,

$$I_0\left(\sqrt{\frac{E}{N_0 T}}\,l_i\right) = \frac{1}{2\pi}\int_{-\pi}^{\pi}\exp\left(\sqrt{\frac{E}{N_0 T}}\,l_i\cos\theta\right)d\theta. \tag{6.79}$$

The likelihood function for signal detection can be expressed in a very compact form as

$$L(s_i) = I_0\left(\sqrt{\frac{E}{N_0 T}}\,l_i\right). \tag{6.80}$$

The binary hypothesis test is now given by

$$I_0\left(\sqrt{\frac{E}{N_0 T}}\,l_1\right) \underset{H_2}{\overset{H_1}{\underset{<}{>}}} I_0\left(\sqrt{\frac{E}{N_0 T}}\,l_2\right). \tag{6.81}$$

It is equivalent to the following test:

$$l_1^2 \underset{H_2}{\overset{H_1}{\underset{<}{>}}} l_2^2. \tag{6.82}$$

A receiver based on the above equation is known as the *quadratic receiver*. Such a test does not rely on other signal-dependent parameter (such as E) and such a test is called *uniformly most powerful* with respect to signal energy E.

We are going to derive two equivalent forms of the quadratic receiver as shown in Figure 6.23(a). Suppose that we have a filter matched to $s(t) = \cos(2\pi f_i t + \theta)$, $0 \le t \le T$. The envelope of a matched filter output is obviously unaffected by θ. Therefore, matched filter can have impulse response $\cos[2\pi f_i (T - t)]$, and its output as the response to received signal $x(t)$ is given by

$$
y(t) = \int_0^T x(\tau) \cos[2\pi f_i (T - t + \tau)] \, d\tau
$$

$$
= \cos[2\pi f_i (T - t)] \int_0^T x(\tau) \cos(2\pi f_i \tau) \, d\tau
$$

$$
- \sin[2\pi f_i (T - t)] \int_0^T x(\tau) \sin(2\pi f_i \tau) \, d\tau. \qquad (6.83)
$$

The envelope of the matched filter output is proportional to the square root of the sum of the squares if the integrals in the above equation. For $t = T$, the envelope is given by

$$
l_i = \left\{ \left[\int_0^T x(\tau) \, \cos(2\pi f_i \tau) \, d\tau \right]^2 + \left[\int_0^T x(\tau) \, \sin(2\pi f_i \tau) \, d\tau \right]^2 \right\}^{1/2}.
$$

$$
(6.84)
$$

We thus obtain a combination of matched filter and envelope detector, which is known as *non-coherent matched filter*.

We are now ready to explore the noise performance of *non-coherent orthogonal modulation*, with two most common cases: non-coherent BFSK and DPSK.

Consider *non-coherent orthogonal modulation* as a binary signaling with two orthogonal and equal-energy signals $s_1(t)$ and $s_2(t)$ (Figure 6.24). Let $g_1(t)$ and $g_2(t)$ denote the phase-shifted version of $s_1(t)$ and $s_2(t)$, which remain orthogonal and equal energy regardless

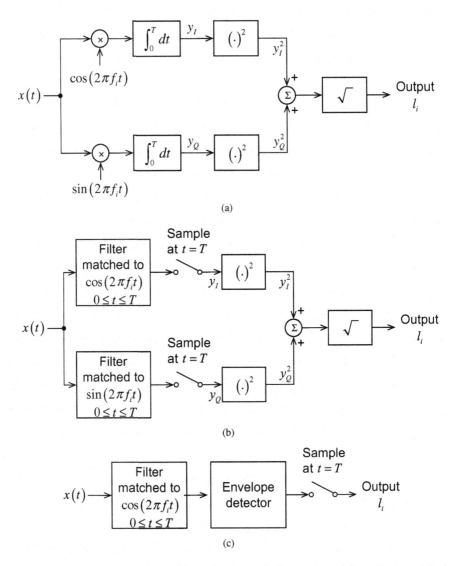

Fig. 6.23 Non-coherent receivers: (a) quadratic correlation receiver; (b) quadratic matched filter; (c) non-coherent matched filter.

of the unknown carrier phase. The channel introduces AWGN $w(t)$ with zero mean and psd $N_0/2$ as follows:

$$x(t) = \begin{cases} g_1(t) + w(t), & s_1(t) \text{ sent}, 0 \le t \le T \\ g_2(t) + w(t), & s_2(t) \text{ sent}, 0 \le t \le T. \end{cases} \tag{6.85}$$

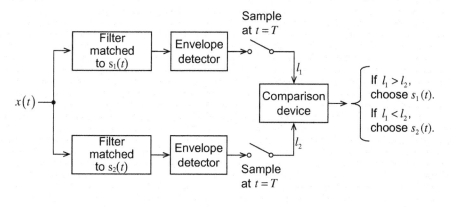

(a)

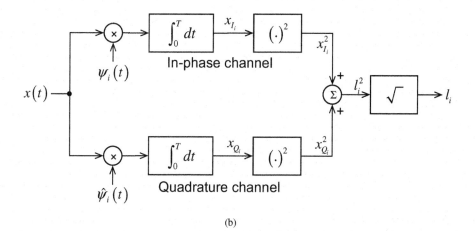

(b)

Fig. 6.24 (a) Generalized binary receiver for non-coherent orthogonal modulation; (b) quadratic receiver.

The quadratic receiver has two channels, in-phase channels, to correlate with

$$\psi_i(t) = m(t)\cos(2\pi f_i t). \tag{6.86}$$

Its Hilbert transform is used for quadrature-phase channel, which is given by

$$\hat{\psi}_i(t) = m(t)\sin(2\pi f_i t). \tag{6.87}$$

Since $\cos\left(2\pi f_i t - \frac{\pi}{2}\right) = \sin(2\pi f_i t)$, the average probability of error for non-coherent orthogonal modulation is expressed as

$$P_e = \frac{1}{2}\exp\left(-\frac{E}{2N_0}\right). \qquad (6.88)$$

For specific case of BFSK, the transmitted signal is given by

$$s_i(t) = \begin{cases} \sqrt{\frac{2E_b}{N_0}}\cos(2\pi f_i t), & 0 \le t \le T_b \\ 0, & \text{elsewhere,} \end{cases} \qquad (6.89)$$

where $f_i = n_i/T_b$ to ensure orthogonality. The non-coherent receiver is as shown in Figure 6.25 and the BER is as in (6.88).

Another type of non-coherent modulation is DPSK, which still requires a coherent reference signal at the receiver but does not need to exactly recover the phase. Its transmitter has two basic operations: (a) differential encoding and (b) PSK. To send "0," we advance current phase of the waveform by 180°. To send "1," we maintain the

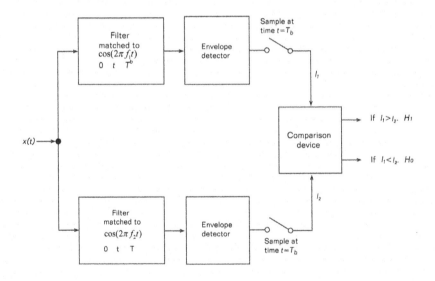

Fig. 6.25 Non-coherent receiver of BFSK.

phase unchanged. If we consider 2 bit intervals, DPSK is a non-coherent orthogonal modulation. For the transmission of "1",

$$
s_1(t) = \begin{cases} \sqrt{\frac{E_b}{2T_b}}\cos(2\pi f_c t), & 0 \le t \le T_b \\ \sqrt{\frac{E_b}{2T_b}}\cos(2\pi f_c t), & T_b \le t \le 2T_b. \end{cases} \tag{6.90}
$$

For the transmission of "0",

$$
s_0(t) = \begin{cases} \sqrt{\frac{E_b}{2T_b}}\cos(2\pi f_c t), & 0 \le t \le T_b \\ \sqrt{\frac{E_b}{2T_b}}\cos(2\pi f_c t + \pi), & T_b \le t \le 2T_b. \end{cases} \tag{6.91}
$$

We can easily observe that $s_1(t)$ and $s_0(t)$ are orthogonal over $[0, 2T_b]$, and thus DPSK is a special case of non-coherent orthogonal modulation with bit-energy as $2E_b$ during $[0, 2T_b]$. From (6.88), the BER of DPSK is given by

$$
P_{e,\text{DPSK}} = \frac{1}{2}e^{-\frac{E_b}{N_0}}. \tag{6.92}
$$

The generation of DPSK is summarized in Table 6.3 and the DPSK transceiver is shown in Figure 6.26.

6.2 Spectral Efficient Modulations

In modern communications, bandwidth is usually important for broadband communications, but very limited (to be considered as bandwidth-limited communications). We intend to transmit as many bits per symbol as possible to increase spectral efficiency. M-ary PSK is obvious by a candidate to meet this goal, though we have to pay price in performance (i.e., BER) degradation.

Table 6.3 DPSK signal.

b_k		1	0	0	1	0	0	1	1
d_{k-1}		1	1	0	1	1	0	1	1
Differentially Encoded d_k	1	1	0	1	1	0	1	1	1
Transmitted Phase	0	0	π	0	0	π	0	0	0

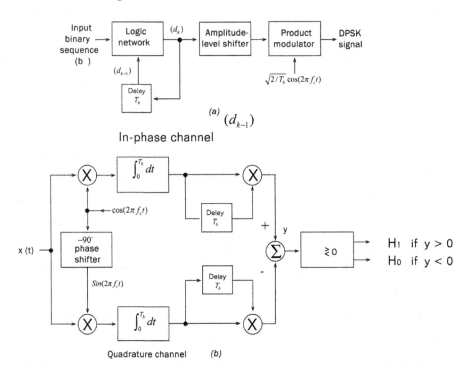

Fig. 6.26 DPSK: (a) transmitter; (b) receiver.

The most common approach is QAM that we introduced earlier. In addition to 16-QAM, we can expand such as 32-QAM, 64-QAM, etc. Signal constellations of 16-QAM, 64-QAM, etc. can be easily created by square-type. Figure 6.27 depicts a general cross-constellation for QAM. For high E_0/N_0, where E_0 is the energy of the lowest signal constellation point, the probability of symbol error is approximately given by

$$P_e \cong 4\left(1 - \frac{1}{\sqrt{2M}}\right)Q\left(\sqrt{\frac{2E_0}{N_0}}\right). \qquad (6.93)$$

Another possible realization is PAM, which is used for digital video broadcasting-satellite version 2. Its signal constellation is shown in Figure 6.28.

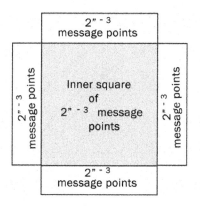

Fig. 6.27 Cross-constellation of QAM.

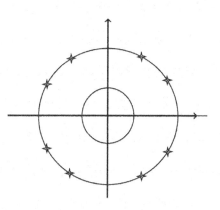

Fig. 6.28 Phase-Amplitude Modulation.

A good example of spectral efficient signaling is voice-band modem, which adopts QAM and error-correcting codes to approach theoretical limit. We will discuss this in the later chapter.

Generally speaking, the selection of modulations is the core to design any digital communications system, which is a trade-off among bandwidth, power, performance (BER), and system complexity. Figure 6.29 summarizes and compares the performance of some PSK and FSK schemes.

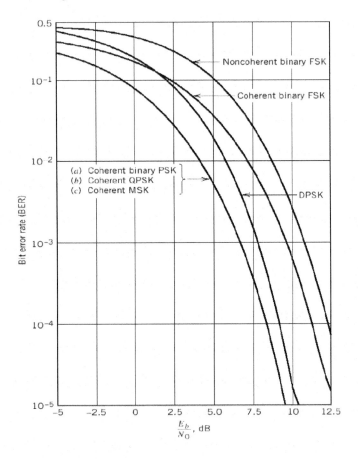

Fig. 6.29 Comparisons of some PSK and FSK modulations.

6.3   Synchronization

Although signal detection/demodulation is the fundamental part of receiver design, a more practically important issue to design a receiver is synchronization to collect carrier (including frequency and phase) and timing information, so that signal detection can successfully execute. Owing to the component variations (such as oscillator imprecision) and propagation/processing delay between the transmitter and the receiver, synchronization to recover such "relative" information is practically needed even if the channel is perfect.

A signal waveform $s(t)$ has been put into a channel. At the receiver, the received waveform, in general, is expressed as

$$r(t) = \alpha(t)e^{j\theta(t)}s(t - \tau) + n(t).$$

To compensate the effect of $\alpha(t)$, we usually apply the *automatic gain control*, which closely works with RF section to ensure a right signal magnitude in receiver processing. To recover $\theta(t)$, it is usually known as the *carrier (phase) recovery*. To get correct τ, it is known as *timing recovery*. In communication theory, we usually treat synchronization as carrier recovery and timing recovery.

Synchronization could be commonly considered into two categories of mechanisms: *data aided* (DA) and *non-DA* (NDA). NDA approach proceeds purely based on the received symbols (or bits), and likely their transition information, while DA approach uses pre-determined pilot signals or preambles as a reference to aid synchronization.

Among various synchronization schemes, we usually have two realizations based on the existence of feedback structure:

- Open loop
- Closed loop

while the closed loops usually have a kind of feedback signal/control. In digital communication systems for networking or broadcasting, a more generalized form can be considered as follows:

$$r(t) = \sum_m \alpha(t)e^{j\theta(t)}s(t - mT - \tau) + n(t).$$

Another dimension of synchronization is therefore needed, known as *frame synchronization* or *block synchronization*, to align bit streams to form a frame or a packet. It usually consists of a unique word with a special autocorrelation property, a header of a packet/frame, payload (information bits) of a packet/frame. A typical unique word is pre-determined and has a correlation peak for perfect alignment of bit streams, while Baker sequences or pseudo-noise sequences are well-known examples.

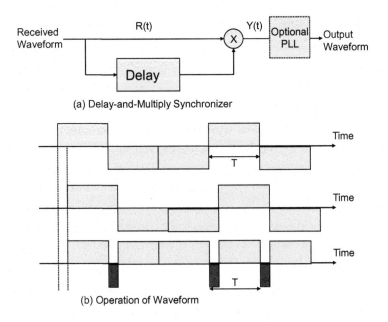

(a) Delay-and-Multiply Synchronizer

(b) Operation of Waveform

Fig. 6.30 Delay-and-multiply type synchronizer.

6.3.1   Open-Loop Spectral Line Generation Methodology

The immediate challenge for receiver prior to signal detection is to acquire appropriate signal parameters such as timing and carrier frequency (that is known as synchronization). The essential task of synchronization is to extract symbol timing from received waveform prior to signal detection/demodulation. Open-loop approach based on waveform correlation operation has been used since analog communications. A good example is delay-and-multiply structure (see Figure 6.30).

6.3.2   Optimal Bit/Symbol Synchronization

We first study the baseband symbol (or bit) synchronization by deriving the optimal scheme. Let the received waveform be expressed as

$$y(t) = s(t, \tau) + n(t) \quad \tau : \text{time jitter}$$

We shall use the Karhünen–Lóeve expansion of a known signal in Gaussian noise to obtain the *maximum a posteriori* estimate of τ, that is,

$$s_k(t,\tau) = \sum_i s_i^{(k)}(\tau)\phi_i(t),$$

where

$$s_i^{(k)}(\tau) = \int_{\tau+(j-1)T}^{\tau+jT} s_k(t,\tau)\phi_i(t)dt;$$

$$\lambda_i\phi_i(\tau) = \int_{\tau+(j-1)T}^{\tau+jT} R_n(t',t)\phi_i(t')dt';$$

$[\tau + (j-1)T, \tau + jT]$ is the observation interval for symbol synchronization;

$$E[n(t)n(t')] = R_n(t,t').$$

It is known that

$$R_n(t,t') = \sum_i \lambda_i\phi_i(t)\phi_i(t')$$

and

$$\int_{\tau+(j-1)T}^{\tau+jT} \phi_i(t)\phi_j(t)dt = \delta_{ij}.$$

If

$$y_N(t) = \sum_{i=1}^{N} y_i\phi_i(t)$$

$$y(t) = \lim_{N\to\infty} y_N(t)$$

then the noise process can be expanded as

$$n(t) = \sum_i n_i\phi_i(t) \quad \text{and} \quad E[n_in_j] = \lambda_i\delta_{ij}.$$

An n-dimensional representation of the observed process over the jth interval is

$$\sum_{i=1}^{N} y_i\phi_i(t) = \sum_{i=1}^{N} s_i^{(k)}(\tau)\phi_i(t) + \sum_{i=1}^{N} n_i\phi_i(t)$$

It is clear that

$$y_i = s_i^{(k)}(\tau) + n_i.$$

Since n_i are independent Gaussians with pdf,

$$P(n_i) = \frac{1}{\sqrt{2\pi\lambda_i}} \exp\left(-\frac{n_i^2}{2\lambda_i}\right)$$

The conditional density of y_i, given τ and binary signal, is expressed as

$$p(y_i|\tau) = \sum_{k=0}^{1} \frac{1}{\sqrt{2\pi\lambda_i}} \exp\left(-\frac{1}{2}\frac{[y_i - S_i^{(k)}(\tau)]^2}{\lambda_i}\right) p(k),$$

where we have averaged over two possible NRZ signals. By $p(k) = \frac{1}{2}, k = 0,1$ (equally probable signaling) we have

$$p(y_1 y_2 \cdots y_N|\tau) = \frac{1}{2}\sum_{k=0}^{1} \left(\frac{1}{2\pi}\right)^{\frac{N}{2}} \prod_{i=1}^{N} \lambda_i^{-\frac{N}{2}} \exp\left(-\frac{1}{2\lambda_i}[y_i - s_i^{(k)}(\tau)]^2\right).$$

For white noise, $\lambda_i = \frac{N_0}{2}, \forall i$, $\mathbf{Q} = \frac{N_0}{2}\mathbf{I}$

$$p(\underline{y}|\tau) = \frac{1}{2}\sum_{k=0}^{1} \left(2\pi\frac{N_0}{2}\right)^{-\frac{N}{2}} \exp\left\{-\frac{1}{2}[\underline{y} - \underline{s}^{(k)}(\tau)]\frac{2}{N_0}\mathbf{I}[\underline{y} - \underline{s}^{(k)}(\tau)]^T\right\}.$$

For antipodal signals such as BPSK,

$$s^{(1)}(t,\tau) = -s^{(0)}(t,\tau) = s(t,\tau)$$

$$p(\underline{y}|\tau) = \sum_{k=0}^{1} \left(2\pi\frac{N_0}{2}\right)^{-\frac{N}{2}} \exp\left(-\frac{1}{N_0}\underline{y}\underline{y}^T\right) \exp\left(-\frac{1}{N_0}\underline{s}_k\underline{s}_k^T\right)$$

$$\times \exp\left(\frac{1}{N_0}\underline{y}\underline{s}_k^T\right) \exp\left(\frac{1}{N_0}\underline{s}_k\underline{y}^T\right)$$

or

$$p(\underline{y}|\tau) = (2\pi)^{-\frac{N}{2}} \left(\frac{2}{N_0}\right)^{\frac{N}{2}} \exp\left(-\frac{1}{N_0}\underline{y}\underline{y}^T\right) \exp\left(-\frac{1}{N_0}\underline{s}\underline{s}^T\right)$$

$$\times \left(\frac{\exp\left(\frac{2}{N_0}\underline{s}\underline{y}^T\right) + \exp\left(-\frac{2}{N_0}\underline{y}\underline{s}^T\right)}{2}\right),$$

where $\underline{s} = \underline{s}_1 = -\underline{s}_2$

$$p(\underline{y}|\tau) = c_1 \cosh\left(\frac{2}{N_0}\underline{y}\underline{s}^T\right).$$

Consequently, $\cosh\left(\frac{2}{N_0}\underline{y}\underline{s}^T\right)$ is the *sufficient statistics*. We know that

$$\frac{2}{N_0}\lim_{N\to\infty}(\underline{y}\underline{s}_1^T) = \frac{2}{N_0}\int_{\tau+(j-1)T}^{\tau+jT} y(t)s^{(1)}(t,\tau)dt \quad \text{successive}$$

For L bits,

$$p(y|\tau) = c_2 \prod_{j=1}^{L} \cosh\left[\frac{2}{N_0}\int_{\tau+(j-1)T}^{\tau+jT} y(t)s^{(1)}(t,\tau)dt\right]$$

By an MAP estimate

$$p(\tau|y(t)) = c_3 \prod_{j=1}^{L} \cosh\left[\frac{2}{N_0}\int_{\tau+(j-1)T}^{\tau+jT} y(t)s^{(1)}(t,\tau)dt\right]$$

$$\Lambda(\tau) = \ln p[\tau|y(t)]$$

$$= \sum_{j=1}^{L} \ln \cosh\left[\frac{2}{N_0}\int_{\tau+(j-1)T}^{\tau+jT} y(t)s^{(1)}(t,\tau)dt\right],$$

where ln cosh is an even function to generate the desired information for synchronization (see Figure 6.31).

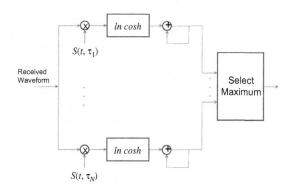

Fig. 6.31 Block diagram of an MAP optimal bit synchronizer.

6.3.3   Optimum Carrier Phase Estimation

We next explore the estimation of carrier phase. Consider a constant amplitude carrier immersed in white Gaussian noise. The received signal is thus given by

$$y(t,\theta) = A\cos(2\pi f_c t + \theta_c) + n(t) \quad 0 \le t \le KT.$$

We intend to obtain the optimal estimate by maximizing the likelihood of $y(t)$ with respect to θ. We first assume θ_c to be a constant. The likelihood function for the unknown phase is given by

$$L(\hat{\theta},\theta) = L[y(t)|\theta = \hat{\theta}]$$

$$= \exp\left\{ \frac{2}{N_0} \int_0^{T_0} [y(t) - A\cos(2\pi f_c t + \hat{\theta})]^2 dt \right\}, \quad (6.94)$$

where T_0 is the observation interval.

We need to consider only the log likelihood to obtain the ML estimate $\hat{\theta}_{\mathrm{ML}}$. By differentiation, we obtain the necessary condition of optimality,

$$\dot{l}(\hat{\theta}) = \int_0^{T_0} [y(t)\sin(2\pi f_c t + \hat{\theta}_{\mathrm{ML}}) + \sin(4\pi f_c t + 2\hat{\theta}_{\mathrm{ML}})]dt = 0.$$

If T_0 is large, then

$$\dot{l}_k(\hat{\theta}) = \int_0^{T_0} y(t)\sin(2\pi f_c t + \hat{\theta}_{ML})dt = 0 \quad T_0 = KT,$$

$$\hat{\theta}_{\mathrm{ML}} = -\tan^{-1} \frac{\int_0^{T_0} y(t)\sin 2\pi f_c t\, dt}{\int_0^{T_0} y(t)\cos 2\pi f_c t\, dt}.$$

The above equations suggest two possible realizations (see Figure 6.32).

6.3.4   Phase Locked Loop (PLL)

The above closed loop realization can be generalized as PLL as shown in Figure 6.33, whose waveform mixer can further be generalized into a phase detector. PLL is widely applied in communication systems and

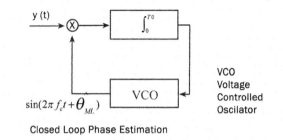

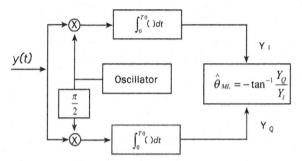

Fig. 6.32 Closed/open-loop phase estimation.

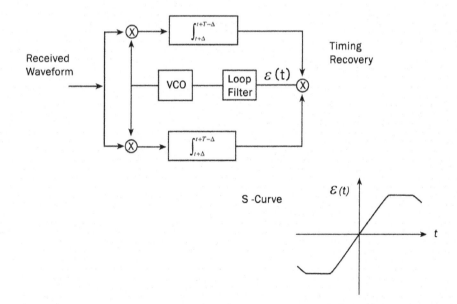

Fig. 6.33 Early-late gate synchronizer and S-curve.

many circuits such as processor and interface control and is described in Chapter 3.

By generalizing the PLL concept, we can develop a sub-optimal timing recovery loop known as *early-late gate synchronizer* that is widely applied in many digital communication systems. Figure 6.33 depicts its structure, which has two arms to advance and to tardy the timing of current estimated waveform. The error signal is then feed into the loop filter to control voltage-controlled oscillator, such that the signal timing can be tracked.

The error signal in fact has a "correlation" nature and we plot as the well-known S-curve. The linear portion of the S-curve implies desired tracking capability for normal operation.

6.3.5   Joint Estimation of Carrier and Symbol Timing

The log likelihood function for these two parameters may be expressed in terms of equivalent low-pass signals as

$$\lambda_L(\phi, \tau) = R_e \left\{ \frac{1}{N_0} \int_{T_0} r(t) v^*(t; \phi, \tau) dt \right\},$$

where $v(t; \phi, \tau)$ is the equivalent low-pass signal, which has the general form:

$$v(t; \phi, \tau) = e^{-j\phi} \left\{ \sum_n I_n u(t - nT - \tau) + j \sum_n J_n w(t - nT - \tau) \right\},$$

where $\{I_n\}$ and $\{J_n\}$ are the two information sequences.

For PAM, $J_n = 0 \ \forall n$, $\{I_n\}$ is real. For QAM and PSK, $J_n = 0 \ \forall n$, $\{I_n\}$ is complex valued. For OQPSK, $\{I_n\}$ and $\{J_n\}$ are non-zero and $w(t) = u\left(t - \frac{T}{2}\right)$. The necessary conditions for the estimates of ϕ and τ to be ML estimates are $\frac{\partial \lambda_L(\phi, \tau)}{\partial \phi} = 0$ and $\frac{\partial \lambda_L(\phi, \tau)}{\partial \tau} = 0$.

6.4   Link Calculations

To design any communication system, especially radio/wireless communication system, we must precede *link calculation* (or *link budget*) at the early stage. It is particularly important in satellite communication (or power-limited communications).

We first decide the required E_b/N_0 to reach certain pre-determined BER. Then, we can define the *link margin* M as

$$\left(\frac{E_b}{N_0}\right)_{\text{rec}} = M \left(\frac{E_b}{N_0}\right)_{\text{req}}$$

Under the same symbol error rate/BER, the link margin would be equivalent to

$$M(\text{dB}) = \left(\frac{E_b}{N_0}\right)_{\text{rec}} (\text{dB}) - \left(\frac{E_b}{N_0}\right)_{\text{req}} (\text{dB})$$

Link margin is useful for enhancing the reliability or availability of a (radio) communication system. In radio communications, we usually first consider a rather ideal free-space propagation model, though we will consider more realistic channels in the later chapter.

We consider the link budget calculations from free-space propagation. Considering a point radiation source with power P_t watts, the power density after distance d is given by

$$\rho(d) = \frac{P_t}{4\pi d^2} \text{ watts/m}^2.$$

The *radiation intensity* measured by watts per unit solid angle is therefore given by

$$\Phi = d^2 \rho(d).$$

Considering a sphere coordinate system, we have the following relationship:

$$d\Omega = \sin\theta \, d\theta \, d\phi \text{ steradians}.$$

The total power is thus expressed as

$$P = \int \Phi(\theta, \phi) \, d\Omega \text{ watts}$$

The average power per unit solid angle can be expressed as

$$
\begin{aligned}
P_{\text{av}} &= \frac{1}{4\pi} \int \Phi(\theta, \phi) \, d\Omega \\
&= \frac{P}{4\pi} \text{ watts/steradian}.
\end{aligned}
$$

The major contribution for transmit antenna and receive antenna in link budget calculations is to better focus radiation and to better collect radiation. The *directional gain* of an antenna is defined as the ratio of the radiation intensity in that direction to the average radiation power. That is,

$$g(\theta,\phi) = \frac{\Phi(\theta,\phi)}{P_{\text{av}}}$$
$$= \frac{\Phi(\theta,\phi)}{P/4\pi}.$$

The *directivity* of the antenna, D, is therefore defined as the ratio of the maximum radiation intensity from the antenna to the radiation from an isotropic source. In other words, D is the maximum of $g(\theta,\phi)$. We can now define the *power gain of an antenna*, G, as the ratio of maximum radiation intensity from the antenna to the radiation intensity from an isotropic radiation source, under the same input power. We usually use η_{antenna} to indicate the *radiation efficiency factor* associated with the antenna, and such a factor is not greater than 1. The power gain of the antenna and the directivity can be linked as follows:

$$G = \eta_{\text{antenna}} D.$$

We are ready to define a well-known term in electromagnetism, *effective isotropic radiation power* (EIRP), for the transmit antenna (with subscript t)

$$\text{EIRP} = P_t G_t \text{ watts.}$$

To quantify the capability of an antenna to transmit/collect radiation power, we introduce a factor known as *effective aperture* of an antenna and it is defined as

$$A = \frac{\lambda^2}{4\pi} G,$$

where λ is the wavelength. The effective aperture is a good measure of efficiency for reflecting antenna such as parabolic antenna in earth station for satellite communications.

We are now ready to formulate equation for free-space radio propagation. We use subscript r to denote receive antenna side (and subscript

t to denote transmit antenna side). The receive power is given by

$$P_r = \left(\frac{\text{EIRP}}{4\pi d^2}\right) A_r$$

$$= \frac{P_t G_t A_r}{4\pi d^2} \text{ watts.}$$

It is equivalent to

$$P_r = P_t G_t G_r \left(\frac{\lambda}{4\pi d}\right)^2.$$

The resulting path loss is expressed as

$$\text{PL} = 10 \log_{10}\left(\frac{P_t}{P_r}\right)$$

$$= -10 \log_{10}(G_t G_r) + 10 \log_{10}\left(\frac{4\pi d}{\lambda}\right)^2.$$

In order to characterize the noise behaviors of components in a system, we use noise figure as the measure. The noise figure, F, is defined under the two-port device model.

The next concept useful in link calculations is the *equivalent noise temperature* that is NOT real temperature. The noise temperature is defined though Boltzmann's constant k and the available noise power is given by

$$N_1 = kT\,\Delta f.$$

For a device, the equivalent noise temperature (again, not the real temperature) is related to the noise figure by

$$T_e = T(F - 1).$$

Now, we consider the cascade connection of two-port networks and the relationship of noise temperature, while G_i denotes the power gain of the ith device with temperature T_i. The overall equivalent noise temperature is given by

$$T_e = T_1 + \frac{T_2}{G_1} + \frac{T_3}{G_1 G_2} + \frac{T_4}{G_1 G_2 G_3} + \cdots.$$

We can easily use a spread sheet program to represent link calculations as the first step to design a (wireless) communication system including all major system components.

6.5   Multiple Access

Up to this moment, we consider all kinds of modulations from a transmitter to a receiver, that is, point-to-point link. However, we usually have multiple radios or terminals in communications, to form a communication network. Recalling the layering structure of computer networks in Chapter 1, we need a sub-layer called medium access control between the data link layer and the physical layer. The purpose of this extra sub-layer is to allocate the multi-access medium various nodes. The method to coordinate physical transmission among various nodes in a computer network is known as multiple access protocol.

The multiple access protocols operate in various network topologies and application scenarios. Figure 6.34(b)(a) is an example of satellite communication networks with earth stations to communication via a satellite, which has a start network topology. Figure 6.34(b)(b) shows a bus network topology while nodes/stations are connected to a network through a bus (typically a cable or a fiber as a physical medium). Another example in Figure 6.34(b)(b) is a multi-hop packet radio network, and this network topology is widely considered in wireless ad hoc networks (some times with a mesh network extension structure) and wireless sensor networks. The world's earliest widely known computer network is a satellite communication network to connect communications in Hawaii islands, with pioneer multiple access protocol ALOHA described in the following section.

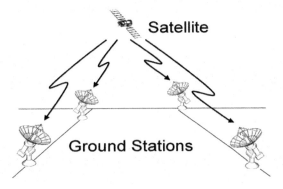

Fig. 6.34(a) Examples of multiple access communication — satellite and ground stations (star network topology).

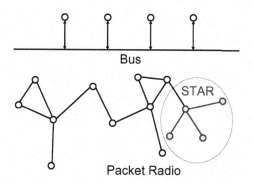

Fig. 6.34(b) Examples of multiple access communication topology — bus, star, and multi-hop packet radio.

6.5.1 Slotted Multi-access and ALOHA System

When a set of nodes share a communication channel, the reception is garbled if two or more nodes transmit simultaneously, which is known as "collision", and the channel is "unused" if none transmit. The challenge of multiple accesses (also multi-access) is how to coordinate the use of such a channel.

Assumptions for idealized slotted multi-access model are summarized as follows:

- Slotted system: All transmitted packets have the same length and each packet requires one time unit (called a slot) for transmission. All transmitters are synchronized such that the reception of each packet starts an integer time and ends at the next integer time.
- Poisson arrival: Packets arrive for transmission at each of the m transmitting nodes according to the independent Poisson processes with λ/m arrival rate.
- Collision or perfect reception.
- {0,1,e} Immediate feedback.
- Retransmission of collisions: Each packet involved in a collision must be retransmitted in some later slot, with further possible such retransmission until successful transmission. A node with retransmission packet is called backlogged.

- No buffer.
- Infinite number of nodes.

6.5.2   Slotted ALOHA

The basic idea of this approach is that each unbacklogged node simply transmits a newly arriving packet in the first slot after the packet arrival, thus risking occasional collisions but achieving very small delay if collisions are rare. For m nodes, an arriving packet in time-division multiplexing (TDM) would have to wait for $m/2$ slots in average to transmit. Slotted ALOHA transmit packets almost immediately with occasional collisions, whereas TDM avoids collisions at the expense of larger delays. When a collision occurs in slotted ALOHA, each node sending one of the colliding packets discovers the collision at the end of the slot and becomes backlogged. Such nodes wait for some random number of slots before retransmitting. With infinite-node assumption, the number of new arrivals transmitted in a slot is a Poisson random variable with parameter λ. If the retransmission from backlogged nodes is sufficiently randomized, it is plausible to approximate the total number of retransmission and new transmissions in a given slot as a Poisson random variable with parameter $G > \lambda$. The parameter of a successful transmission in a slot is Ge^{-G} Figure 6.35. The maximum achievable rate called throughput is therefore $1/e$.

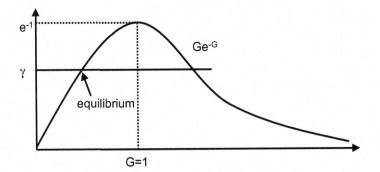

Fig. 6.35 Throughput of Slotted-Aloha.

6.5.3 Unslotted ALOHA

There is no slotted concept in pure ALOHA that can be considered as the most primitive version of multiple access protocols Figure 6.36. A station with a packet to transmit just transmits and listens to the channel. If a packet is involved in a collision, it is retransmitted after a random delay.

We may observe from Figure 6.37 that the collided duration is up to two times greater than that of slotted ALOHA. Therefore,

$$\text{Throughput} = G(n)e^{-2G(n)}.$$

The maximum achievable rate is thus $\frac{1}{2e}$.

Regarding further delay analysis of ALOHA family protocols and more multiple access protocols, readers may explore more in literatures or any textbook in computer networks or data networks.

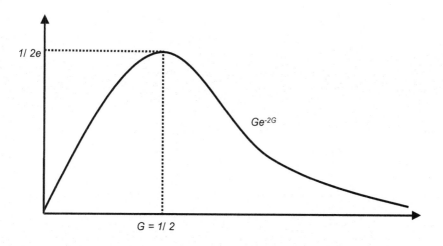

Fig. 6.36 Unslotted ALOHA throughput.

Fig. 6.37 Packets in a collision.

6.6   Exercises

1. If we want to transmit W bps signaling, answer the following questions:

   (a) If we are using BFSK-CP, what is the minimum bandwidth? Draw by assuming a carrier frequency f_c. Hint: MSK.

   (b) Find the coherent optimum receiver and its associated BER, under AWGN.

   (c) Repeat (b) for non-coherent demodulation.

   (d) If we are using a raised-cosine waveform shaping with roll-off factor 25%, repeat (a). What is the reason to conduct waveform shaping?

   (e) If we are using four-level FSK-CP, can we have better (in terms of BER and spectral efficiency) signaling performance?

2. Consider the following equally probable digital modulation with symbol period T: When "1" is transmitted, $A sin(2\pi f_1 t)$ is received. When "0" is transmitted, $-A sin(2\pi f_2 t)$ is received.

   (a) What is the condition to yield the highest possible spectral efficiency under this type of signaling?

   (b) Under AWGN with zero mean and two-side psd $N_0/2$ observed at the receiver input, find the BER (or error probability) in terms of E_b/N_0 using coherent demodulation.

   (c) Repeat (b) for non-coherent demodulation.

   (d) Derive the optimal timing recovery scheme for such a binary signaling.

   (e) If we are using four-level FSK-CP, can we have better (in terms of BER and spectral efficiency) signaling performance?

   (f) Describe how to treat MSK as a sort of FSK, and what is the optimal demodulation in this case.

3. Consider the following digital modulation: When "1" is transmitted, $A sin(2\pi f_c t)$ is received. When "0" is transmitted, $-A sin(2\pi f_c t)$ is received.

   (a) Under AWGN with zero mean and two-side psd $N_0/2$ observed at the receiver input, find the BER (or error probability) in terms of E_b/N_0.

   (b) If there exists an phase error θ to recover the above signal waveform at the receiver, find the BER if θ is fixed, deterministic, and unknown.

   (c) [Bonus] If θ is random and uniformly distributed in $(0,2\pi)$, repeat (a).

   (d) How can we improve the spectral efficiency for this modulation without losing error rate performance? Also find the symbol error rate and BER relationship in this case.

4. Consider the following digital modulation: When "1" is transmitted, $A sin(2\pi f_c t)$ is received. When "0" is transmitted, $-A sin(2\pi f_c t)$ is received.

   (a) Under AWGN with zero mean and two-side psd $N_0/2$ observed at the receiver input, find the BER (or error probability) in terms of E_b/N_0.

   (b) If the signal experiences a fading to result in varying amplitude (i.e. A) and the pdf of A is $f_A(a)$, $a \geq 0$, find the average BER. Hint: Chapter 8.

5. Consider a set of M orthogonal signal waveforms $s_{m(t)}$, $1 \leq m \leq M$, $0 \leq t \leq T$, all of which have the same energy ε. Define a new set of M waveforms as

$$s'_m(t) = s_m(t) - \frac{1}{M} \sum_{k=1}^{M} s_k(t), \quad 1 \leq m \leq M, \ 0 \leq t \leq T$$

Show that the M signal waveform $\{s'_m(t)\}$ have equal energy, given by

$$\varepsilon' = (M-1)\varepsilon/M$$

and are equally correlated, with correlation coefficient

$$\gamma_{mn} = \frac{1}{\varepsilon'} \int_0^T s_m'(t)s_n'(t)dt = -\frac{1}{M-1}.$$

6. A binary digital communication system employs the signals

$$s_0(t) = 0, \quad 0 \le t \le T$$

$$s_1(t) = A, \quad 0 \le t \le T$$

for transmitting the information. This is called ON–OFF signaling. The demodulator crosscorrelates the received signal $r(t)$ with $s_1(t)$ and samples the output of the correlator at $t = \mathrm{T}$.

(a) Determine the optimum detector for an AWGN channel and the optimum threshold, assuming that the signals are equally probable.

(b) Determine the probability of error as a function of the *SNR*. How does ON–OFF signaling compare with antipodal signaling?

7. A carrier component is transmitted on the quadrature carrier in a communication system that transmits information via BPSK. Hence, the received signal has the form:

$$v(t) = \pm\sqrt{2P_s}\,\cos(2\pi f_c t + \phi) + \sqrt{2P_c}\,\sin(2\pi f_c t + \phi) + n(t),$$

where ϕ is the carrier phase and $n(t)$ is AWGN. The unmodulated carrier component is used as a pilot signal at the receiver to estimate the carrier phase.

(a) Sketch a block diagram of the receiver, including the carrier-phase estimator.

(b) Illustrate mathematically the operations involved in the estimation of the carrier-phase ϕ.

(c) Express the probability of error for the detection of the BPSK signal as a function of the total transmitted power $P_T = P_s + P_c$. What is the loss in performance due to the allocation of a portion of the transmitted power to the pilot signal? Evaluate the loss for $P_c/P_T = 0.1$.

8. Consider the PLL for estimating the carrier phase of a signal in which the loop filter is specified as

$$G(s) = \frac{K}{1 + \tau_1 s}.$$

   (a) Determine the closed-loop transfer function $H(s)$ and its gain at $f = 0$.

   (b) For what range of value of τ_1 and K is the loop stable?

9. A radio communication system transmits at a power level of 0.1 watt at 1 GHz. The transmitting and receiving antennas are parabolic, each having a diameter of 1 m. The receiver is located 30 km from the transmitter.

   (a) Determine the gains of the transmitting and receiving antennas.

   (b) Determine the EIRP of the transmitted signal.

   (c) Determine the signal power from the receiving antenna.

10. Binary antipodal signals are used to transmit information over an AWGN channel. The prior probabilities for the two input symbols (bits) are 1/3 and 2/3.

    (a) Determine the optimum maximum-likelihood decision rule for the detector.

    (b) Determine the average probability of error as a function of ε_b/N_0.

11. A baseband digital communication system employs the signals shown in the following figure (a) for transmission of two equiprobable messages. It is assumed that the communication problem studied here is a "one shot" communication problem, that is, the above messages are transmitted just once and no transmission takes place later. The channel has no attenuation ($\alpha = 1$) and the noise is AWG with power spectral density $\frac{N_0}{2}$.

    (a) Find an appropriate orthonormal basis for the representation of the signals.

(b) In a block diagram, give the precise specifications of the optimal receiver using matched filters. Label the block diagram carefully.

(c) Find the error probability of the optimal receiver.

(d) Show that the optimal receiver can be implemented by using just one filter [see the block diagram shown in the following figure(b)]. What are the characteristics of the matched filter and the sampler and decision device?

(e) Now assume that the channel is not ideal, but has an impulse response of $c(t) = \delta(t) + \frac{1}{2}\delta(t - \frac{T}{2})$. Using the same matched filter used in the previous part, design an optimal receiver.

(f) Assuming that the channel impulse response is $c(t) = \delta(t) + a\delta(t - \frac{T}{2})$, where a is a random variable uniformly distributed on $[0,1]$, and using the same matched filter, design the optimal receiver.

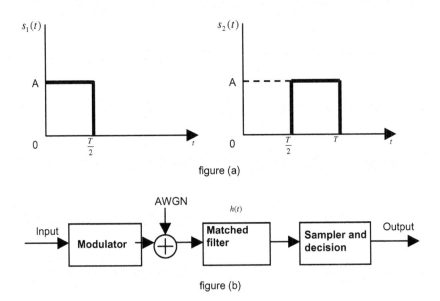

figure (a)

figure (b)

7

Error-Correcting Codes

As mentioned in Chapter 1, channel capacity indicates the maximal transmission rate to achieve reliable communications. However, there is no specific indication about the way to achieve this goal. One major advantage of digital communication is to utilize digital techniques to control errors caused by the channel. In fact, we commonly use two kinds of technologies to achieve error control in digital communication: automatic repeat request (ARQ) and forward error-correcting codes (FEC). ARQ primarily controls the errors at packet (or frame) level. FEC relies on adding redundant bits into transmission, which can protect information against noise and undesirable channel effects. We shall introduce the concepts of both ARQ and FEC in this chapter.

7.1 ARQ

ARQ controls packet (i.e. a frame of bits) transmission errors between a transmitter–receiver (Tx–Rx) pair with a feedback channel indicating erroneous received packets. Whether the packet/frame is correct is usually through the so-called *cyclic redundancy check* (CRC) adopting codes introduced in Section 7.2. The most intuitive ARQ is *stop-and-wait ARQ*, which functions as follows:

(1) Tx sends a packet to Rx. Rx runs the CRC check and then sends a positive acknowledgement (P-ACK) to Tx through the feedback channel, for a correct reception; or, Rx sends a negative acknowledgement (N-ACK) to Tx for an erroneous reception.

(2) In case Rx receives N-ACK, Rx re-transmits the packet again. In case Rx received P-ACK, it proceeds transmission for the next packet.

Such a mechanism is reliable. However, stop-and-wait ARQ has drawbacks in wasting communication bandwidth by waiting and some memory to store the packets. In today's broadband communication, stop-and-wait ARQ is hardly a good choice. For communication bandwidth efficiency, *selective ARQ* can be employed as the following algorithm:

(1) Tx keeps transmission of packets and Rx keeps reception. Rx sends an N-ACK with sequence number back to Tx, if Rx CRC check identifies an error.

(2) Once N-ACK with sequence number is received at Tx, Rx transmits the indicated packet again by inserting into the packet stream.

Under great bandwidth efficiency, keen readers might note an obvious price to pay in selective ARQ, the huge memory to store a good number of packets. Although the memory cost keeps dropping quickly, the tremendously increasing amount of information keeps this concern even more seriously. A good compromise between selective ARQ and stop-and-wait ARQ is the *go-back-N* ARQ, which is widely used not only in communications but also in various data link control protocols in computer networks. Among several variations, we use Figure 7.1 as a typical example to explain the operation of go-back-N ARQ, where $N = 4$. For a successful reception of a packet, after updating the request number to Tx, the sliding window has been updated from [0,3] to [1,4]. For an erroneous reception of a packet and after time-out, the whole frame of packets 1–4 is transmitted again. In this manner, we keep a balance between bandwidth efficiency and memory size. A great deal of investigations about optimal operating parameters in different environments have been conducted in the past years.

7.2 Linear Block Codes

The immediate idea to introduce redundant bits to protect information is via checksum. Let us start from the following definitions. In this chapter, we consider only binary bits in "1" or "0."

Definition 7.1. An (n, k) block code has a length of n bits, with $k < n$ information bits. Therefore, there are 2^k binary sequences of length n,

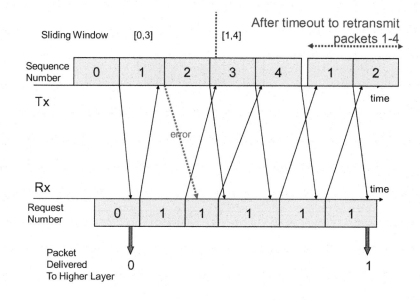

Fig. 7.1 Go-back-4 ARQ.

and each sequence is called a *code word*. A code $\mathbf{C}$ consists of 2^k code words and $\mathbf{C} = \{\mathbf{c_1}, \mathbf{c_2}, \dots, \mathbf{c_{2^k}}\}$.

Definition 7.2. The *Hamming distance* between two code words $\mathbf{c_i}$ and $\mathbf{c_j}$, $d_H(\mathbf{c_i}, \mathbf{c_j})$, is the number of components that two code words differ.

Definition 7.3. The minimum distance of a code is the minimum Hamming distance between any two code words. That is,

$$d_{\min} = \min_{\substack{\mathbf{c_i}, \mathbf{c_j} \\ i \neq j}} d(\mathbf{c_i}, \mathbf{c_j}).$$

Definition 7.4. A block code is *linear* if any linear combination of two code words is another code word. In binary cases, if $\mathbf{c_i}$ and $\mathbf{c_j}$ are code words, $\mathbf{c_i} \oplus \mathbf{c_j}$ is a code word, where $\oplus$ denotes component-wise modulo-2 addition.

An (n, k) linear block code is usually generated by a $k \times n$ *generator matrix* as follows:

$$
\mathbf{G} = \begin{bmatrix} g_{11} & g_{12} & \cdots & g_{1n} \\ g_{21} & g_{22} & \cdots & g_{2n} \\ \vdots & \vdots & \ddots & \vdots \\ g_{k1} & g_{k2} & \cdots & g_{kn} \end{bmatrix}.
\tag{7.1}
$$

For a k-bit data vector $\mathbf{d} = (d_1, d_2, \ldots, d_k)$, the code word is

$$
\mathbf{c} = \mathbf{dG}.
\tag{7.2}
$$

Example 7.1.   A (5,2) code is defined as $C = \{00000, 10100, 01111, 11011\}$. The code is linear. We can map the information bits into code words as follows:

$$
\begin{aligned}
00 &\to 00000 \\
01 &\to 01111 \\
10 &\to 10100 \\
11 &\to 11011
\end{aligned}
$$

The generator matrix (derived from information sequences 10 and 01) is therefore given by

$$
\mathbf{G} = \begin{bmatrix} 10100 \\ 01111 \end{bmatrix}.
\tag{7.3}
$$

Note a feature of (7.3) from the view of parity check. We may interpret (7.3) as

$$
(c_1, c_2, c_3, c_4, c_5) = (x_1, x_2)\mathbf{G}.
\tag{7.4}
$$

That is,

$$
\begin{aligned}
c_1 &= x_1 \\
c_2 &= x_2 \\
c_3 &= x_1 \oplus x_2 \\
c_4 &= x_2 \\
c_5 &= x_2.
\end{aligned}
\tag{7.5}
$$

The first two coded bits are just information bits and others are extra parity checks. Such a code is called a *systematic code*.

Proposition 7.5. A necessary and sufficient condition of a code being systematic is the generator matrix of the form:

$$\mathbf{G} = \begin{bmatrix} \mathbf{I_k} \vdots \mathbf{P} \end{bmatrix}, \tag{7.6}$$

where $\mathbf{I_k}$ is a $k \times k$ identify matrix and $\mathbf{P}$ is a $k \times (n-k)$ binary matrix to generate parity check.

The decoding of the systematic linear block codes replies on the parity check matrix defined as follows:

$$\mathbf{H} = \begin{bmatrix} -\mathbf{P^t} \vdots \mathbf{I_k} \end{bmatrix}. \tag{7.7}$$

Note that $-\mathbf{P^t} = \mathbf{P^t}$ for binary cases.

Example 7.2. Following the code in Example 7.1, the parity check matrix is given by

$$\mathbf{H} = \begin{bmatrix} 11 \vdots 100 \\ 01 \vdots 010 \\ 01 \vdots 001 \end{bmatrix}. \tag{7.8}$$

We can generate the standard array as follows. Each column is known as a *coset* and its first element is known as a *coset leader*:

$$\begin{matrix} \mathbf{c_1} & \mathbf{c_2} & \cdots & \mathbf{c_{2k}} \\ \mathbf{e_1} & \mathbf{e_1} \oplus \mathbf{c_2} & \cdots & \mathbf{e_1} \oplus \mathbf{c_{2k}} \\ \vdots & \vdots & \ddots & \vdots \\ \mathbf{e_{2^{n-k}-1}} & \mathbf{e_{2^{n-k}-1}} \oplus \mathbf{c_2} & \cdots & \mathbf{e_{2^{n-k}-1}} \oplus \mathbf{c_{2k}}. \end{matrix} \tag{7.9}$$

Suppose $\mathbf{y}$ to be the observed binary sequence from detection. The decoding algorithm that is suitable for hardware implementation proceeds as follows:

(a) Calculate the syndrome of $\mathbf{y}$ as $\mathbf{s} = \mathbf{y}\mathbf{H}^t$.
(b) Using the standard array, identify the coset corresponding to $\mathbf{s}$.
(c) Obtain the coset leader $\mathbf{e}$ and decode $\mathbf{y}$ as $\mathbf{c} = \mathbf{y} \oplus \mathbf{e}$.

Proposition 7.6 (Hamming Codes). The well-known Hamming codes are a class of linear block codes with $n = 2^m - 1$, $k = 2^m - m - 1$, and $d_{\min} = 3$.

Example 7.3 A popular (7,4) Hamming code has generator matrix as

$$\mathbf{G} = \begin{bmatrix} 1000 \vdots 110 \\ 0100 \vdots 011 \\ 0010 \vdots 101 \\ 0001 \vdots 111 \end{bmatrix}.$$

A subset of linear block codes is cyclic codes, with an extra condition that if $\mathbf{c}$ is a code word, then its cyclic shift(s) is also a code word. Encoder and decoder of cyclic codes are usually easy to implement. The Famous BCH codes and Reed–Solomon codes belong to this class. Interested readers may find a standard textbook of error-correcting codes (such as the one written by S. Lin and D. Costello) for further reading. In recent years, low-density parity check (LDPC) codes that were invented in 1960s as a kind of block codes became an attractive approach to Shannon limit, whose details can be found in many books.

7.3   Convolutional Codes

A popular class of FEC without nice mathematical property as linear block codes but with good performance is convolutional codes. Different

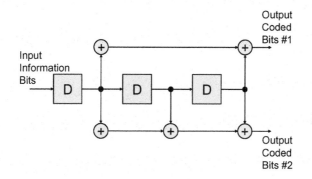

Fig. 7.2 Convolutional encoder.

from block codes, encoding of convolutional codes maps k information bits into n bits to transmit over the channel but these n bits are determined by current information bits and also previous information bits. The code rate r is thus expressed as

$$r = \frac{k}{n}. \tag{7.10}$$

To briefly introduce the concept of convolutional codes, we illustrate the following example with $k = 1$, $n = 2$, and $K = 3$, where K stands for the *constraint length* of the code. Figure 7.2 depicts the encoder for $r = 1/2$, $K = 3(K$ shift registers) convolutional codes.

The corresponding generator sequences are given by

$$\begin{aligned} \mathbf{g_1} &= [1\ 0\ 1] \\ \mathbf{g_2} &= [1\ 1\ 1]. \end{aligned} \tag{7.11}$$

Or, we can express in polynomials as

$$\begin{aligned} g_1(D) &= 1 + D^2 \\ g_2(D) &= 1 + D + D^2. \end{aligned} \tag{7.12}$$

We can clearly observe the memory property of convolutional codes, which also suggests using a *state-transition* diagram to represent the behaviors of the convolutional code. Figure 7.3 shows such a state-transition.

Based on the state-transition diagram, a more popular method of more insight to describe convolutional codes is to specify the corresponding *trellis diagram*. Trellis diagram can show transitions between

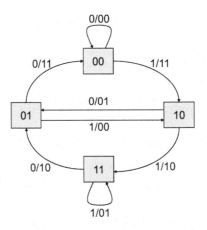

Fig. 7.3 State-transition diagram.

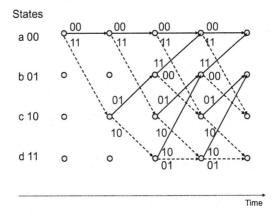

Fig. 7.4 Trellis diagram.

states as time evolves. Figure 7.4 depicts the trellis diagram for the convolutional codes in Figure 7.2. Note that convolutional codes always start from the state of all zeros, that is, 00 in this example. In fact, the code should end at the state of all zeros (i.e., 00 in this example).

The beauty of trellis is its capability of graphical description for the code. Considering an information sequence (1100), the encoder translates the sequence into (11101011) as shown in Figure 7.5 (in red wide line).

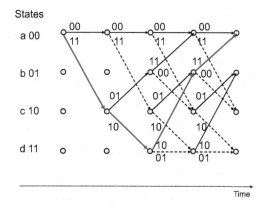

States

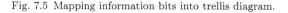

Fig. 7.5 Mapping information bits into trellis diagram.

The above nature suggests the optimal decoding of convolutional codes to find a path in trellis diagram that has minimum distance from a given sequence. This is the well-known *Viterbi algorithm*, which can be used not only in such decoding, but also in maximum likelihood sequence estimation, and many others requiring optimal trellis searching.

In Viterbi algorithm, the optimal path connecting $S_1 = 0$ (starting from all-zero state) to $S_{i-1} = v$ is called a survivor path at state $S_{i-1} = v$. Defining $\delta(S_1 = 0, S_i = l)$ as the distance metric, we can update the new survivor metric as

$$\delta(S_1 = 0, S_i = l) = \min_{v} \{\delta(S_1 = 0, S_{i-1} = v) + \delta(S_{i-1} = v, S_i = l)\}. \tag{7.13}$$

Proposition 7.7 (Viterbi Algorithm). The procedure can be summarized as follows.

(i) Parse the received sequence into m sub-sequences of length n.

(ii) Draw a trellis of depth m for the code.

(iii) Set $i = 1$ and set the metric of initial all-zero state equal to 0.

(iv) Find the distance of the ith sub-sequence of the received sequence to all branches connecting states at the ith stage to states at the $(i + 1)$th stage.

(v) Add these distances to the metrics of states at the ith stage to obtain the metric candidates for states at the $(i + 1)$th stage.

(vi) For each state at the $(i + 1)$th stage, select the candidate(s) of minimum metric and label the branch corresponding to this minimum value as the survivor and assign the candidate(s) as the metric of states at the $(i + 1)$th stage.

(vii) If $i = m$, go to step (viii). Otherwise, increase I by 1 and go to (iv).

(viii) Start with the all-zero state and go back through the trellis along the survivors to reach the initial all-zero state. This path is the optimal path and the corresponding input-bit sequence is the maximum likelihood decoded information sequence.

In case the distance of Viterbi algorithm is Hamming distance, it is known as hard decision Viterbi algorithm. In case we use the demodulator output to form the sequence to be considered, and we use Enclidean distance, such an algorithm is known as soft decision, as the optimal decoding of convolutional codes.

Finally, modern communication systems and digital broadcast systems are likely to adaptively use multiple-rate coding with multiple-rate modulation. To simplify decoder design for multiple-rate convolutional codes, puncturing is introduced. Suppose that we have a rate 1/3 convolutional encoder and decoder. We may encode 2 information bits into 6 bits. By transmitting these 6 bits, we can decode accordingly by rate 1/3 decoder. To switch into rate 1/2 coding, we may simply puncture 2 coded bits from 6 bits and transmit these 4 bits. By randomly adding redundant 2 bits back, we may use the same rate 1/3 decoder to complete rate 1/2 decoding. Instead of separate 2 decoders, we can use one with almost no penalty as long as we select and arrange properly.

7.4 Trellis-Coded Modulation

Spectral efficient modulation and coding play a critical role for high-bandwidth communications. We usually have to use high spectral efficient modulation such as quadrature amplitude modulation (QAM). However, such high spectral efficient modulations are usually sensitive to various kinds of errors. One major factor is phase jump to create a series of subsequent errors in the demodulation of symbols.

QAM signal constellation (or any constellation with 90° rotation symmetry) can be resistant to phase jumps by a differential coding of data bits in quadrants. Any sudden change of phase in channel can result in only single symbol error without further error propagation. Otherwise, errors may occur during the recovery of phase synchronization over several symbol intervals. For example, (b_1, b_2, b_3, b_4) determining a signal point in 16 QAM, (b_1, b_2) determines the quadrant and are differentially encoded.

When $(0, 1, 1, 1)$ are coded into S_n, a phase jump of 90° has occurred in the transmission channel (of course, it is unlikely to be exactly 90°). There is a symbol error from the phase jump Figure 7.6. It is desirable to implement Gray coding in the mapping bits to signal points, which

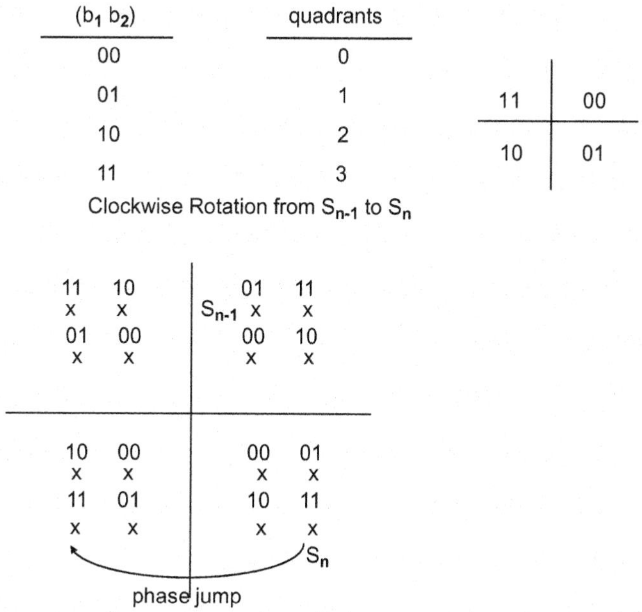
(b1 b2) quadrants
00 0
01 1
10 2
11 3
Clockwise Rotation from Sn-1 to Sn
11 00
10 01
11 10 01 11
x x Sn-1 x x
01 00 00 10
x x x x
10 00 00 01
x x x x
11 01 10 11
x x x x
Sn
phase jump

Fig. 7.6 Phase jump in QAM.

minimize the number of bits in error when a symbol error is made due to noise.

After the successful employment of high spectral efficient modulation, we may adopt forward error correction code (FEC) to help sensitivity to channel errors. The introduction of FEC can increase the minimum distance between data sequences by introducing redundancy. When the expanded sequence is transmitted through a noisy, distorted, band-limited channel, a loss of performance must be weighed against the coding gains. The loss may be either in distortion caused by expanding the signal bandwidth into less desirable channel characteristics or by expanding signal constellation, which keeps the same bandwidth but reduces the noise margin between signal points under the same average power constraint. However, for decades, communication engineers always treat modulation and coding as separate subjects in system design.

In 1976, Ungerboeck found that substantial gains could be achieved in telephone channels by convolutional coding of the signal levels rather than the source data. The unified coding and modulation can directly increase the "free distance" (minimum Euclidean distance) between coded line signal sequences. The greater the Euclidean distance between signal sequences, the smaller the error rate, which is approximated by

$$P_e = N_{\text{free}} \cdot Q\left(\frac{d_{\text{free}}}{2\sigma}\right) \quad \text{for large } SNR, \tag{7.14}$$

where N_{free} is the average number of nearest-neighbors at distance d_{free}.

With "codewords" consisting of modulation level, not all sequences can be codewords, thus a gain in minimum Euclidean separation of sequence over minimum distance between signal points.

CCITT (now ITU-T) recommendations for 9.6, 14.4, 19.2 Kbps modem have all adopted TCM. Ungerboeck noted that finding codes with good Euclidean distance does *NOT* follow the knowledge to find codes with Hamming distance, which is the conventional way to implement channel (convolutional) coding. Figure 7.7 demonstrates a simple 4-state $r = 2/3$ trellis code for 8-PSK. By properly mapping codewords, the Euclidean distance can be increased from 8-PSK minimum separation to that for QPSK, which may dramatically enhance resistance

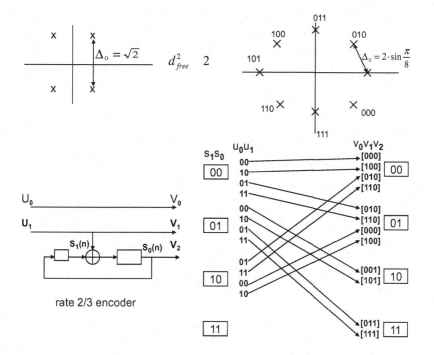

Fig. 7.7 Ungerboeck's 4-state Trellis codes for 8-PSK

against errors. 8-PSK constellation shall be partitioned so that parallel branches in the trellis represent signal points at maximum separation, as illustrated in Figure 7.8.

The total squared Euclidean distance between these two paths is given by

$$d^2 = \left(d_{\text{leavry the same mode}}\right)^2 + \left(d_{\text{entery different modes}}\right)^2$$

$$+ \left(d_{\text{entery the same mode}}\right)^2 > 2 + \left(2\sin\frac{\pi}{8}\right)^2 + 2 > 4. \quad (7.15)$$

Consequently, we can use the Euclidean distance that is directly related to error rate performance in additive white Gaussian noise (AWGN) (recall the meaning of Equation (5.175) that $d_{\min}$ dominates error rate performance in AWGN general signaling), to define the *coding gain* as follows:

$$Coding\ Gain = 10\log_{10}\left[\frac{d_{\min}^2(coded\ 8\text{-}PSK)}{d_{\min}^2(uncoded\ QPSK)}\right] \cong 3\,\text{dB}. \quad (7.16)$$

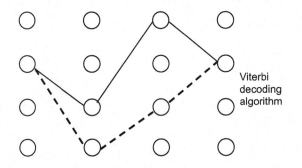

Viterbi
decoding
algorithm

Fig. 7.8 Trellis illustration in Viterbi algorithm.

Note that such a way of calculating coding gain is true only when the *SNR* is large. For conventional convolutional coding with 8-PSK to achieve similar performance of this example, we require much more complex codes, i.e., much more constraint length and thus states. To generalize such an idea, Ungerboeck introduced the *set partitioning principle* to increase the Euclidean distance between signal sequences.

As shown in Figure 7.9, the general TCM encoding model for channel trellis coding with m information bits, $m' < m$, of these bits is applied to a rate $m'/m' + 1$ conventional encoder, generating $m' + 1$ code bits. These bits are used to select one of $2^{m'+1}$ subsets from a redundant $2^{m'+1}$-element signal set. Set partitioning divides a signal constellation successively into smaller constellations with increasing at

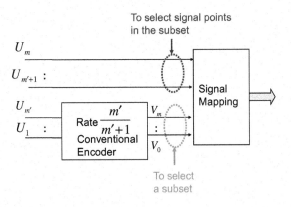

Fig. 7.9 General TCM encoder.

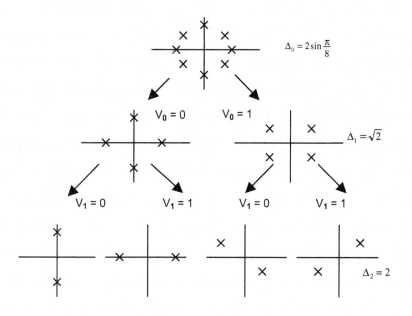

Fig. 7.10 Set partitioning for 8-PSK.

least an intraset distance Δ_i, $i = 0, 1, \ldots$. The partitioning is repeated $m' + 1$ times until $\Delta_{m'+1}$ is at least as large as the desired free distance of the trellis code. In Figure 7.10, 8-PSK can proceed into signal constellation of distance 2, instead of originally $2 \sin \pi/8$, which can enhance coding performance due to the increase in Euclidean distance for signal constellation.

The free Euclidean distance of a channel trellis code is given by

$$d_{\text{free}} = \min\{\Delta_{m'+1}, d_{\text{free}}(m')\} \qquad (7.17)$$

Note that TCM works well in Gaussian noise channel, but is sensitive to phase offset. Rotational-invariant TCM is thus desirable. V.33 (14.4 Kbps data tx. over 4-wire circuits) adopted a nonlinear convolutional coding discovered by L.F. Wei to use an 8-state code invariant to $90°$ rotations with 4 dB gain over linear code. The fundamental principle is that the convolutional coder translates a sequence of differentially coded data into a stream of $90°$ or its multiples occur; all received subsequent symbols are incremented by the same amount of jump. The subsequent convolutional decoding also results in a data stream that, although the data stream fed into the decoder is not the same as the original, has the original differential coding and can therefore be translated into the correct data. TCM is widely applied for many bandwidth efficient communication systems.

In 1990s, turbo codes using multiple convolutional codes with sequential decoding algorithm further approach Shannon limit, except the coding delay problem remaining a challenge.

8

Communications Over Wireless in Fading Channels

Owing to the booming of applications of wireless communications, modern communication system design is likely related to such scenarios. State-of-the-art wireless communications usually propagate signals over *fading channels*, likely caused by multipath effects. Based on previous chapters, we shall introduce more topics in depth regarding communication over fading channels.

8.1 Channel Modeling

Modern wireless broadband communications supporting high-bandwidth communications almost always experience non-ideal channel effects such as fading and channel distortion. In order to design a successful physical transmission system over wireless medium, we must be able to appropriately model the channel according to its measured characteristics. Based on the channel model(s), we can further design and analyze the radio transmission link to develop desired functional blocks.

From the early days when people applied radio for communication at high-frequency (HF) band, fading is a serious problem for communication engineers. In radio communications, fading is typically caused by the multipath effects. Signal after transmission over multiple-path radio propagation may have several signal components from the same signal source but having different amplitudes, phases, and even frequencies. The receiver combines all such signal components corresponding to different paths together and results in constructive and destructive interferences from such superposition. A destructive interference can cause signal envelope dropping quickly in extremely short time duration up to several tens of dBs from its regular level, which is known as a *fade*, and such a phenomenon is called *fading*. Fading can result in error rate dramatically increasing and/or system *outage*. Such a performance bottleneck is usually the major focus in our system design.

Radio propagation over channels involves basic mechanisms including free-space propagation, reflection, diffraction, and scattering (as shown in Figure 8.1) to induce multipath effects to cause rapid changes in envelope and phase, etc. Factors to influence fading include multipath propagation, Doppler speed caused by the terminal (either transmitter or receiver), moving of surrounding subjects, transmission bandwidth, etc. Radio propagation and thus fading is very difficult to predict, and we usually model the channel in a statistical manner based on tremendous measurements. Figure 8.2 is a popular and the simplest example to model mobile radio channel for cellular communications, two-ray fading by sending an impulse from the transmitter to the receiver through a channel.

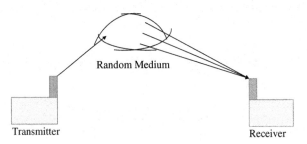

Random Medium

Transmitter

Receiver

Fig. 8.1 Multipath fading.

Mathematically, a multipath fading channel is usually modeled as a linear filter with a complex-valued impulse response, which can be expressed as

$$h(t) = \sum_{k=0}^{N-1} a_k \delta(t - \tau_k) e^{j\theta_k}, \tag{8.1}$$

where N is the number of multipath components; $\{a_k\}$, $\{\tau_k\}$, and $\{\theta_k\}$ are the random amplitude, propagation delay, and phase sequences, associated with these paths respectively; and δ is the delta function. Static channels are completely characterized by these path variables.

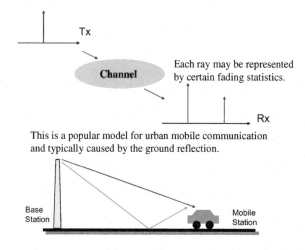

Fig. 8.2 Two-ray fading model.

It is easy to obtain the response $y(t)$ of the channel when any signal $s(t)$ is transmitted by convolving $s(t)$ with $h(t)$ and adding noise, that is

$$y(t) = \int_{-\infty}^{\infty} s(t)h(t - \tau)d\tau + n(t), \qquad (8.2)$$

where $n(t)$ is the low-pass, complex-value additive Gaussian noise. This is the basic frame-structure to proceed link analysis. For some channels such as tropospheric scattering, continuous form may be more suitable.

If we model the channel impulse response as a zero-mean complex-valued Gaussian process, we call this channel as the *Rayleigh fading channel* if the envelope$|h(t)|$ is Rayleigh-distributed, and as the *Ricean fading channel* if the envelope has a Rician distribution. The squared-envelope of Rayleigh fading, with α^2 as average power, is distributed as

$$f_{A^2}(x) = \frac{1}{\alpha^2}e^{-\frac{x}{\alpha^2}}. \qquad (8.3)$$

Similarly, the squared-envelope of Ricean fading, with K as the ratio of the power in the specular and scattered components, has a distribution as

$$f_{A^2}(x) = \frac{K+1}{\alpha^2}e^{-K-\frac{(K+1)x}{\alpha^2}}I_0\left(2\sqrt{\frac{K(K+1)x}{\alpha^2}}\right). \qquad (8.4)$$

We may intuitively consider Rayleigh fading as a lot of small signal components without a major signal component, which is usually a direct path, and consider Rician fading as a major signal with a lot of smaller signal components. In other words, Rayleigh may represent a worse fading situation than Rician. Both fading statistics have random phase of uniform distribution.

In fact, the multipath effects spread the signal pulse not only in time domain but also in frequency domain to cause the so-called intersymbol interference. Under the assumption of uncorrelated scattering (i.e. the attenuation and phase shift associated with path τ_1 is uncorrelated with the attenuation and phase shift associated with path τ_2), if we ignore the frequency domain effect, the resulting complex autocorrelation function is simply the average power output of the channel as a function of time delay τ. This is known as the *multipath intensity profile* or the *delay power spectrum*. The range of τ where non-zero multipath intensity profile is called the multipath *spread* of the channel and is denoted by T_m. We define the *coherence bandwidth* of the channel as $1/T_m$. If the transmitted signal bandwidth is greater than the coherence bandwidth, then this channel is said to be *frequency selective*. Otherwise, it is *non-frequency selective*. Similarly, the time variations in the channel are considered as a Doppler spread. Its Fourier transform is the *Doppler power spectrum* of the channel. The inverse of non-zero range of Doppler power spectrum can be similarly defined as the *coherence time*. If the coherence time is greater than the symbol period, then it is called *flat fading*. Otherwise, it is called *fast fading*. A slowly changing channel has a large coherence time, or equivalently a small Doppler spread.

If we want to completely model the mobile radio channel, we need to consider the following factors to form long-term fading and short-term fading:

- Shadowing;
- Path-loss;
- Fast fading;
- Correlation among sub-bands.

In mobile communications, log-normal channel modeling is usually adopted for long-term shadow fading, while Rayleigh and Rician channel models are usually considered for short-term channel models. Details and models for more realistic fading channels can be found in many mobile communications textbooks such as *Principles of Mobile Communications* by G. Stuber, etc.

8.2 Bit-error-rate (BER) Analysis Over Fading Channels

Fading harms communication link between the transmitter and the receiver in two ways: by resulting signal unexpected fades up to several tens of dBs from normal value, which usually invoke system outage; by damaging BER. Figure 8.3 depicts an example of computer generating fading along the time progressing, while we can easily observe the fades more than 20 dB once a while. Since we commonly model unpredicted fading in terms of statistics, we usually consider *average BER* to evaluate system performance under fading channels.

Let us consider BER analysis for binary phase shift keying (BPSK) under a Rayleigh fading channel with additive white Gaussian noise

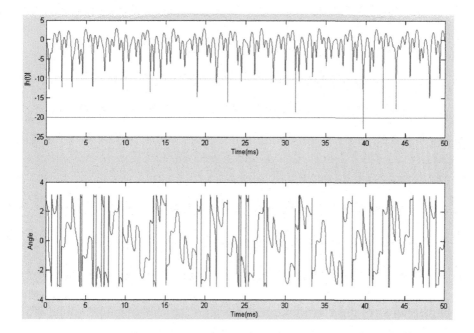

Fig. 8.3 Fading channel (100 km moving speed).

(AWGN). Recall the BER of BPSK in AWGN, $P_{e,\text{BPSK}} = Q\left(2\frac{E_b}{N_0}\right)$. The average BER for BPSK in AWGN and Rayleigh fading is obtained using

$$\overline{P_e} = \int_0^\infty P_{e,\text{BPSK}|x} f_{\text{Rayleigh}}(x)\, dx. \tag{8.5}$$

Using (8.3), we have

$$\overline{P_e} = \int_0^\infty Q\left(2x\frac{E_b}{N_0}\right)\frac{1}{\alpha^2} e^{-\frac{x}{\alpha^2}}\, dx$$

$$= \frac{1}{2}\left(1 - \sqrt{\frac{\gamma_0}{1+\gamma_0}}\right), \tag{8.6}$$

where $\gamma_0 = \alpha^2 \frac{E_b}{N_0}$. For coherent binary frequency shift keying (FSK) and binary differential PSK (DPSK) in Rayleigh fading, the respective average BERs are given by

$$\bar{P}_{e,\text{BFSK}-\text{Coherent}} = \frac{1}{2}\left(1 - \sqrt{\frac{\gamma_0}{2+\gamma_0}}\right);$$

$$\bar{P}_{e,\text{DPSK}} = \frac{1}{2(1+\gamma_0)}.$$

8.3 Diversity

As we can see fading creating a great challenge in wireless communications, there are several common ways to combat with fading in this performance bottleneck of system design, in addition to meaningless way by increasing transmission power:

- Equalization;
- FEC;
- Diversity.

We have introduced the principles of equalization and forward error correcting code (FEC) in previous chapters. However, diversity might be the most useful technique to overcome the challenge of fading. The fundamental idea of diversity is to create independent transmission channels to receive and to demodulate signal suffering from fading.

There are a few means to create such independent channels:

- *Time diversity*: The signal is transmitted to the receiver at different times, with the simplest example to repeat transmission for a number of times.

- *Frequency diversity*: The signal can be transmitted through multiple frequency (two or more) channels simultaneously or at different times.

- *Polarization diversity*: The same signal can be transmitted through different polarizations (right and left) of electromagnetic waves, or I-channel and Q-channel.

- *Code diversity*: The same signal can be transmitted by using different coding (either forward error correcting code (FEC) or spreading codes of spread spectrum communications described in Chapter 10).

- *Antenna diversity*: Diversity usually requires extra consumption of resources, either in time domain or in frequency domain, to reduce the spectral efficiency. We may use multiple (two or more) antennas at the receiver to collect the transmitted signal. Provided antennas are properly separated (more than half wave-length in theory, and practically more), the receiver can still receive from "independent" channels, which is commonly known as *receive diversity* and introduces only slightly more system complexity. By appropriate selecting antenna(s) and combining received waveforms, we can expect substantial diversity gain in system performance. The same multiple-antenna principle can be extended to the transmitter end and is known as *transmit diversity*. Simultaneous utilization of transmit diversity and receive diversity formulates the well-known *multiple-input-multiple-output (MIMO)* processing of signals to significantly enhance link quality. Figure 8.4 illustrates spatial multiplexing as one of the well-known MIMO techniques.

- *Space diversity and/or macro diversity*: Diversity receptions can be realized at much larger distance by putting sensors/receivers geographically separated, which is used in code

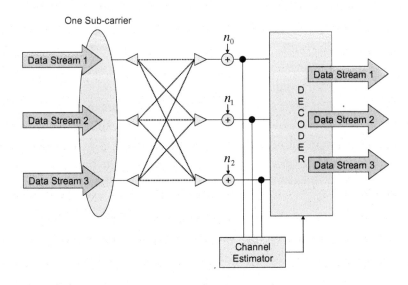

Fig. 8.4 Spatial multiplexing.

division multiple access (CDMA) cellular. We may consider cooperative communication concept originating from this point.

To introduce the analysis of diversity, we start from receive diversity. Prior to demodulation, the receiver has to combine waveforms from diversity, which is known as *combining*. The most straightforward combining is *equal-gain* combining, by equally weighted all received waveforms from different channels (or antennas). We may clearly observe a drawback from such equal-gain combining, that is, any weak signal-to-noise (SNR) channel/waveform/branch can be the performance bottleneck. Therefore, *selective combining* is widely used for receive diversity by selecting the branch with the highest SNR, especially for wireless packet transmission. In what follows, we will analyze its BER performance under Rayleigh fading. Note that the primary purpose to use diversity is to combat with fading.

Rayleigh fading experienced in the kth branch of selective diversity, with mean power α^2, is distributed as

$$f_{\gamma_k}(x) = \frac{1}{\alpha^2} \cdot e^{-x/\alpha^2}.$$

The selective combing always chooses the largest (instantaneous) SNR (or E_b/N_0) among totally L branches:

$$\gamma_{sc} = \max\{\gamma_1, \gamma_2, \gamma_3, \ldots, \gamma_L\}.$$

To calculate probability density function of the selective combining, we may apply cumulative distribution functions of ordered statistics as follows:

$$F_{\gamma_{sc}}(x) = \Pr\left\{\bigcap_{l=1}^{L} \gamma_l \leq x\right\}$$

$$= (1 - e^{-x/\alpha^2})^L.$$

The second equality comes from independence among L branches, that is,

$$f_{\gamma_{sc}}(x) = \frac{L}{\alpha^2} \cdot (1 - e^{-x/\alpha^2})^{L-1} \cdot e^{-x/\alpha^2}.$$

The average SNR (or E_b/N_0) for SC is given by

$$\bar{\gamma}_{sc} = \int_0^{\infty} x \cdot f_{\gamma_{sc}}(x) \cdot dx$$

$$= \alpha^2 \cdot \sum_{l=1}^{L} \frac{1}{l}.$$

We can therefore use to calculate the average BER using selective diversity in the fading channel. For example, DPSK with differential detection has BER

$$P_{e,\text{DPSK}} = \frac{1}{2} \cdot e^{-SNR}.$$

With selective diversity, the average BER under Rayleigh fading is given by

$$\bar{P}_{e,\text{DPSK-SC}} = \int_0^{\infty} P_{e,\text{DPSK}}(x) \cdot f_{\gamma_{sc}}(x)\, dx$$

$$= \int_0^{\infty} \frac{L}{2\alpha^2} \cdot e^{-(1+\frac{1}{\alpha^2})} \cdot (1 - e^{-x/\alpha^2})^{L-1}\, dx$$

$$= \frac{L}{2} \cdot \sum_{l=0}^{L-1} \frac{\binom{L-1}{l} \cdot (-1)^l}{1 + l + \alpha^2}.$$

Keen readers might be curious about the optimal combining based on the maximum likelihood (ML) criterion, which is known as *maximum ratio combining* (MRC):

$$\underline{\omega}_l = g_l \underline{\mathbf{s}}_m + \underline{\mathbf{n}}_m \qquad l = 1, \ldots, L$$
$$\underline{\omega} \equiv (\underline{\omega}_1, \underline{\omega}_2, \ldots, \underline{\omega}_L) \quad m = 1, \ldots, M$$
$$g_l = a_l \cdot e^{-j\phi_l},$$

which has a multivariate Gaussian distribution as

$$P(\underline{\omega}|\underline{g}, \underline{\mathbf{s}}_m) = \frac{1}{(\pi N_0)^{LM}} \cdot e^{-\frac{1}{N_0} \cdot \sum_{l=1}^{L} \|\underline{\omega}_l - g_l \underline{\mathbf{s}}_m\|^2},$$

where $\underline{g} = (g_1, g_2, g_3, \ldots, g_L)$ is the channel vector. The ML receiver selects the message vector $\underline{\mathbf{s}}_m$ that maximizes the metric

$$\mu(\underline{\mathbf{s}}_m) = -\sum_{l=1}^{L} \|\underline{\omega}_l - g_l \underline{\mathbf{s}}_m\|^2.$$

The same principle is used as discussed in Chapter 5, with equally probable and equal-power signaling,

$$\mu(\underline{\mathbf{s}}_m) = \sum_{l=1}^{L} \mathrm{Re}\{\underline{\omega}_l \cdot g_l^* \underline{\mathbf{s}}_m^*\}$$

$$= \mathrm{Re}\left\{\left(\sum_{l=1}^{L} g_l^* \cdot \underline{\omega}_l\right) \cdot \underline{\mathbf{s}}_m^*\right\}$$

The difference for the second equality lies in the fact that the weighting and combining is performed before the integration. The envelope of the composite signal is given by

$$a_M = A \cdot \sum_{l=1}^{L} a_l^2.$$

The sum of the branch noise powers, if P_n is the noise power in each diversity branch, is expressed as

$$P_{n,\text{total}} = P_n \cdot \sum_{l=1}^{L} a_l^2.$$

The *SNR* (or E_b/N_0) with MRC is given by

$$\gamma_{\text{MRC}} = \frac{a_M^2}{2 \cdot P_{n,\text{total}}} = \sum_{l=1}^{L} \gamma_l.$$

If all diversity branches have the same average power and are uncorrelated, then γ_{MRC} has a chi-square distribution of $2L$ degrees of freedom, that is,

$$f_{\gamma_{\text{MRC}}}(x) = \frac{1}{(L-1)!(\alpha^2)^L} \cdot x^{L-1} \cdot e^{-x/\alpha^2}.$$

The average *SNR* with MRC is given by

$$\bar{\gamma}_{\text{MRC}} = \sum_{l=1}^{L} \gamma_l = L \cdot \alpha^2.$$

Note that the above equation can be derived based on Cauchy–Schwartz inequality too. Since MRC is based on coherent detection, we can use it only with coherent modulations. For example, the average BER of BPSK in Rayleigh fading, after some algebraic manipulations, is given by

$$\bar{P}_{e,\text{BPSK–MRC}} = \int_0^{\infty} Q(\sqrt{2x}) \cdot f_{\gamma_{\text{MRC}}}(x)\, dx$$

$$= \left(\frac{1-\mu}{2}\right)^L \cdot \sum_{l=0}^{L-1} \binom{L-1+l}{l} \left(\frac{1+\mu}{2}\right)^l,$$

where $\mu = \sqrt{\frac{\alpha^2}{1+\alpha^2}}$.

The MRC set the foundation of many further technologies such as MIMO, RAKE receiver, etc.

9

Orthogonal Frequency Division Multiplexing

Up to this moment, the communication technologies that we have discussed are using single carrier. In this chapter, we are going to consider multi-carrier communication technology.

9.1 Introduction

In early satellite communications, communications based on different carrier frequencies may go through one transponder in a communication satellite. The earth station may also combine signals with different carrier frequencies through one power amplifier to satellite, or through one low-noise amplifier from satellite. This can be considered as a sort of multi-carrier communications. When a new communication challenge to provide highly bandwidth efficient transmission for asymmetric digital subscriber line arises, J. Cioffi et al. developed multi-carrier technology by dividing available spectrum over the wire into many frequency sub-bands and by adjusting modulation constellations over these sub-bands based on channel situation. State-of-the-art wireless communications adopt a similar concept as multi-carrier technology known as orthogonal frequency division multiplexing (OFDM). As mentioned in Chapter 1, OFDM and its multiuser version OFDMA have been widely used in the IEEE 802.11 g/a wireless local area networks (also known as WiFi) and IEEE 802.16e (also known as mobile WiMAX), 3G LTE, etc., state-of-the-art wireless broadband communications.

The origin of OFDM could be traced back to R. Chang by using a set of parallel transmitters and receivers as shown in Figure 9.1. The fundamental idea behind this is to carry signals separately on several orthogonal "sub-carrier" frequencies, for example, the nth sub-carrier

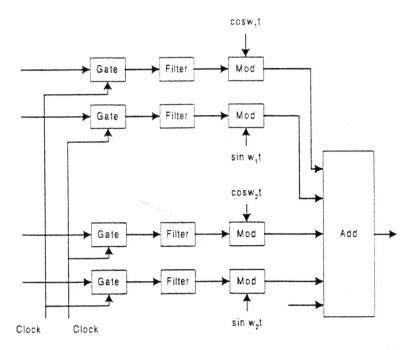

Fig. 9.1 Original concept of OFDM.

having frequency at nf_b, $n = 1, 2, 3, \ldots, N$. These sub-carrier sinusoidal waveforms are orthogonal, and good spectral efficiency is expected.

However, one major drawback in Figure 9.1 is its great complexity to prohibit practical interests to apply. S. Weinstein and P. Ebert then proposed using discrete Fourier transform (DFT) to overcome this problem to reduce implementation to one transmitter and one receiver. Migrating from Fourier transform in Chapter 2 to deal with frequency domain analysis of discrete-time signals and systems, we can define N-point DFT as follows.

Definition 9.1. The N-point DFT of a finite length sequence $x(n)$ is given by

$$X(k) = \sum_{n=0}^{N-1} x(n)e^{-j2\pi kn/N}. \tag{9.1}$$

The inverse transform called inverse DFT (IDFT) provides a way to recover the finite length sequence as

$$x(n) = \frac{1}{N} \sum_{k=0}^{N-1} X(k)e^{j2\pi kn/N}. \tag{9.2}$$

Calculations of DFT/IDFT can be relatively large in modern communication systems. Fast Fourier transform (FFT) for $N = 2^n$, $n = 1, 2, 3, \ldots$, has therefore proposed to resolve this challenge and has been used for practical realization of OFDM. Figure 9.2 illustrates the principle of eight-point FFT operation by defining $W_N = e^{-j2\pi/N}$ with $W_N^{k+N/2} = -W_N^k$, which forms butterfly to significantly reduce calculations. The number of required multiplications (that implies the computation complexity) reduces from the order of N^2 to $N \log N$. Observing from Figure 9.2, if we represent the index of $x(n)$ and $X(k)$ by 3 binary bits, their orders in time domain and frequency domain are just reversed, for example, input $x(100)$ and output $X(001)$ in the second line. IFFT in fact proceeds as FFT just by considering replacement of index. In modern OFDM systems, we usually adopt FFT/IFFT.

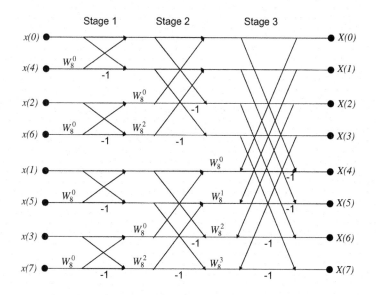

Fig. 9.2 FFT.

However, OFDM at this stage still suffers from delay spread in fading channels due to the loss of orthogonal property among sub-carriers resulted from fading delay to create *inter-symbol interference* (ISI) and *inter-channel interference* (ICI), where ICI means interference from adjacent sub-carrier(s). This challenge is brilliantly resolved by adding a *cyclic prefix* (or zero-padding) during the FFT, which is equivalent to creating a guard-band to avoid ISI and ICI. Figure 9.3 depicts OFDM with cyclic extension. Len Cimini proved this technique to be useful for OFDM wireless systems in fading channels to open a new era.

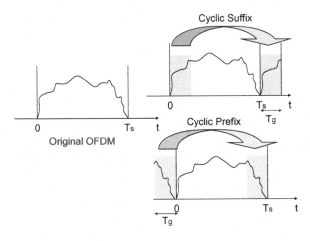

Fig. 9.3 OFDM with cyclic extension.

The original OFDM concept can be realized as follows. The transmitted signal spectrum is chosen so that ICI does not occur. More precisely, the spectrum of each sub-carrier (sub-channel) is zero at other sub-carrier frequencies. We modulate N data in parallel, spaced by $\Delta t = 1/f_s$ (f_s is the symbol rate) with N sub-carrier frequencies. The signaling interval T increases to $N\Delta t$, which suggests the system more vulnerable to delay spread from fading and channel impairments. As shown in Figure 9.4, the transmitted waveform is given by

$$D(t) = \sum_{n=0}^{N-1} \{a(n)\cos\omega_n t + b(n)\sin\omega_n t\}, \qquad (9.3)$$

where $\omega_n = 2\pi f_n$, $f_n = f_0 + n\Delta f$, and $\Delta f = \frac{1}{N\Delta t}$.

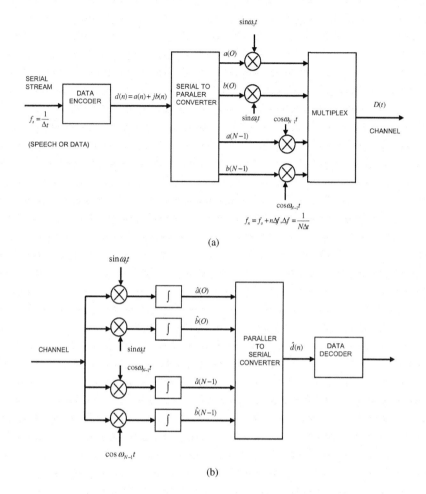

Fig. 9.4 Basic OFDM: (a) transmitter; (b) receiver.

To explore the spectral efficiency of OFDM, we suppose that the symbol rate of serial data is $1/\Delta t$. The bit rate for a corresponding M-ary system is $\log_2 \frac{M}{\Delta t}$. Each sub-channel transmits at a much lower rate $\log_2 \frac{M}{N\Delta t}$ due to N parallel sub-channels. The total bandwidth of the OFDM system is given by

$$B = f_{N-1} - f_0 + 2\delta, \tag{9.4}$$

where f_n is the frequency of the nth sub-carrier, and δ is the one-side bandwidth of the sub-channel. The sub-carriers are orthogonally packed and uniformly spaced so that $f_{N-1} - f_0 = (N-1)\Delta f$. Since $\Delta f = \frac{1}{N\Delta t}$ and $\delta = \frac{\Delta f}{2} = \frac{1}{2N\Delta t}$, $f_{N-1} - f_0 = (1 - \frac{1}{N})\frac{1}{\Delta t}$. As we practically reserve some extra guard for filtering by a factor of α (i.e., $\delta = (1 + \alpha)\frac{1}{2N\Delta t}$), the spectral efficiency η is given by

$$\eta = \frac{\log_2 M}{\left(1 - \frac{1}{N}\right) + 2\delta\Delta t} = \frac{\log_2 M}{1 + \frac{\alpha}{N}} \leq \log_2 M. \tag{9.5}$$

To reach the highest spectral efficiency in OFDM, N shall be large (i.e., more sub-carriers) and α shall be small.

Now, we are ready to implement OFDM through DFT, for communication over fading channels, and the ways to deal with major challenges of OFDM (i.e., system complexity and severe interference among sub-channels due to channel distortion. We consider the implementation of OFDM system as shown in Figure 9.5 to explain its principles.

From (9.3), such an implementation suggests that

$$D(t) = \text{Re}\left[\sum_{n=0}^{N-1} d(n)e^{-j\omega_n t}\right]. \tag{9.6}$$

Channel fading and possible co-channel interference may distort the signal to impair signal orthogonality among sub-carriers. The fading

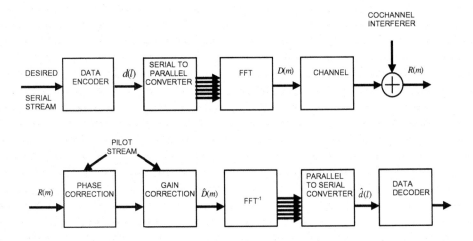

Fig. 9.5 OFDM system implementation via FFT.

can be modeled as

$$Z(m) = A(m)e^{j\theta(m)}, \tag{9.7}$$

where $A(m)$ are contributed from fading statistics (such as Rayleigh) and $\theta(m)$ are from a uniformly distributed random phase. The output data sequence is given by

$$
\begin{aligned}
\hat{d}(l) &= \frac{1}{N}\sum_{m=0}^{N-1} Z(m)D(m)e^{j\frac{2\pi}{N}lm} \\
&= \sum_{m=0}^{N-1} d(n)\left[\frac{1}{N}\sum_{m=0}^{N-1} Z(m)e^{j\frac{2\pi}{N}m(l-n)}\right] \\
&= \sum_{n=o}^{N-1} d(n)z(l-n), \tag{9.8}
\end{aligned}
$$

where $z(l)$ is the IDFT of $Z(m)$. Equation (9.8) implies a complex average of samples of the complex fading envelope. Without any fading, $Z(m) = 1$, and $z(l - n) = \delta(l - n)$; thus $\hat{d}(l) = d(l)$. In the presence of fading,

$$\hat{d}(l) = d(l)z(0) + \sum_{n=0,n\neq l}^{N-1} d(n)z(l-n). \tag{9.9}$$

The second term means that the ICI and ISI resulted from loss of orthogonality. In fact, ICI can create great harm to OFDM communications. To ensure good communication, pilot signals are usually inserted into transmission to serve as the reference of amplitude and carrier phase to aid the performance of a receiver.

Let $D(m)$ be the desired signal sequence, and $I(m)$ be the interference. With

$$
\begin{aligned}
Z_d(m) &= A_d(m)e^{j\theta_d(m)} \\
Z_i(m) &= A_i(m)e^{j\theta_i(m)}
\end{aligned}
$$

the sequence present at the receiver is represented as

$$R(m) = Z_d(m)D(m) + \sqrt{\gamma}Z_i(m)I(m),$$

where $\sqrt{\gamma} = SIR^{-1}$. $R(m)$ is corrected by a complex sequence $Z_p(m)$, and the complex pilot fading envelope gives

$$\hat{D}(m) = \frac{Z_d(m)}{Z_p(m)} D(m) + \sqrt{\gamma} \frac{Z_i(m)}{Z_p(m)} I(m).$$

If unlimited gain and phase correction is used, there is no ISI and we can obtain

$$\hat{d}(l) = d(l) + \frac{\sqrt{\gamma}}{N} \sum_{m=0}^{N-1} I(m) \frac{Z_i(m)}{Z_d(m)} e^{j \frac{2\pi}{N} ml}.$$

The only distortion is caused by co-channel interference.

Modern OFDM systems usually have a block diagram as shown in Figure 9.6. Owing to the duality between time domain and frequency domain in FFT or IFFT, many functional blocks can be realized in time domain or in frequency domain (that is, prior to or after FFT/IFFT). Figure 9.6 is just a typical example in IEEE 802 type wireless communication systems. Blocks in the light gray (yellow) area are usually implemented in analog circuits. We may also note that IFFT is used in the transmitter and FFT is used in the receiver, as the common notations in literatures.

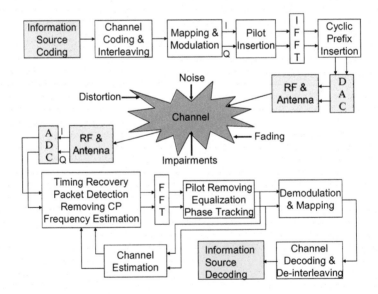

Fig. 9.6 Typical block diagram of OFDM systems.

9.2 Synchronization for OFDM Signals

The fundamental open-loop synchronization is to create the correlation between the received signal and its delayed waveform. Such a correlation has information at the signal symbol rate (or zero-crossing information). We may further use a properly selected phase-locked loop (PLL) to lock the desired signal information to achieve the purpose of synchronization for timing or carrier recovery:

$$X(m) = D(m)e^{j2\pi f_c mT_s}$$

$$r(m) = D(m)e^{j2\pi(\Delta f)mT_s}$$

Let Δf be the frequency error and $\delta$to be the delay to generate discrimination:

$$\Delta = \sum_{n=0}^{N-1} r_n r_{n+\delta}^*$$

$$= \sum_{m=0}^{N-1} D(m)e^{j2\pi\Delta fmT_s}[D(m+\delta)e^{j2\pi\Delta f(m+\delta)T_s}]^*$$

$$= \left[\sum_{m=0}^{N-1} D(m)D^*(m+\delta)\right]e^{-j2\pi\Delta f\delta T_s}.$$

Since $\sum_{m=0}^{N-1} D(m)D^*(m+\delta) = \sum_{m=0}^{N-1}|D(m)|^2$, we can obtain the estimate of carrier as

$$\Delta\hat{f} = -\frac{1}{2\pi\delta T_s}\arg(\Delta).$$

By similar concepts, OFDM synchronization can therefore generally derived for the timing error as $\varepsilon = \frac{T_0}{T_s/N}$ and angle error (while we have two portions as integer frequency and fractional phase) as $\frac{\delta f}{\Delta f} = k_0 + \theta$. Let λ be the channel length in the number of sample periods. Timing offset Estimation can be proceeded as

$$y_n = x_n + n_n$$

$$= \sum_{k=0}^{N-1} H_k S_k e^{j2\pi\frac{(k+k_0-\theta)(n-N_g-\varepsilon)}{N}}$$

$$\begin{cases} = \sum_{k=0}^{N-1} H_k S_k e^{j2\pi \frac{(k+k_0-\theta)(n-N_g-\varepsilon)}{N}} + n_n + \text{ISI} + \text{ICI} & n \in [\varepsilon, \varepsilon + \lambda] \\ = \sum_{k=0}^{N-1} H_k S_k e^{j2\pi \frac{(k+k_0-\theta)(n-N_g-\varepsilon)}{N}} + n_n & n \in [\varepsilon + \lambda, \varepsilon + N + N_g]. \end{cases}$$

Here we can easily identify ISI from the previous OFDM symbol and ICI from misaligned samples in the current OFDM symbol:

$$\hat{\varepsilon}_{\text{ML}} = \arg \max_{\varepsilon} \Lambda_{\text{ML}}(\tau)$$

$$= \arg \max_{\varepsilon} \sum_{n=N_g+\varepsilon}^{N_g+\varepsilon+\frac{N}{2}-1} \left\{ \text{Re} \left[y_n^* y_{n+\frac{N}{2}} \right] - \frac{\zeta}{2} \left(|y_n|^2 + |y_{n+\frac{N}{2}}|^2 \right) \right\},$$

where $\zeta = \frac{SNR}{SNR+1}$.

Regarding frequency offset estimation, we can proceed as

$$y_n = \sum_{k=0}^{N-1} H_k S_k e^{j2\pi \frac{(k+k_0+\theta)n}{N}}$$

$$= X_n e^{j2\pi \frac{(k_0+\theta)n}{N}} \quad |\theta| < \frac{1}{2}.$$

Such a task is composed of two stages: fine frequency estimate θ and coarse frequency estimate k_0. For a non-pilot-based system,

$$\hat{\theta} = \frac{1}{2\pi} \arg \left(\sum_{n-0}^{N-1} y_n^* y_{n+N} \right).$$

For a system with pilot signals,

$$\hat{\theta} = \arg \left(\sum_{n=\varepsilon}^{\varepsilon+N_g-1} y_n y_{n+N}^* \right).$$

We can also precede joint time–frequency estimation together by the following concept. It is equivalent to find the following optimization jointly with timing and (carrier) phase:

$$(\hat{\tau}_{n+1}, \hat{\varphi}_{n+1}) = \underset{\substack{\hat{\tau}_n - \Delta\tau, \hat{\tau}_n, \hat{\tau}_n + \Delta\tau \\ \hat{\varphi}_n - \Delta\varphi, \hat{\varphi}_n, \hat{\varphi}_n + \Delta\varphi}}{\arg} |s(\tau, \varphi)^* s(\hat{\tau}, \hat{\varphi})|.$$

Such an optimization can be facilitated by different recursive algorithms.

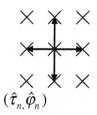

$(\hat{\tau}_n, \hat{\varphi}_n)$

Fig. 9.7 Current estimator and 8 neighboring candidates.

Finally, a very important factor to design an OFDM system is the *Phase Noise*. There are many ways to invoke phase noise in OFDM system, that is,

$$x_n = \sum_i \text{IDFT}\{s_{ik}\} g_n(n - iN),$$

where $g(\cdot)$ represents the pulse function with a rectangular shape of N samples of duration.

Modulating signal: zero-mean white Gaussian random process $\varphi_s(n)$ with variance σ_s^2. Its autocorrelation is given by

$$R_{\varphi_s}(k) = \sigma_s^2 \cdot \delta(k)$$

$$R_{\varphi_s}(f) = \sum_{k=\infty}^{\infty} R_{\varphi_s}(k) \cdot e^{-j2\pi fk} = \sigma_s^2.$$

After low-pass filter (LPF) with impulse response $h_{\text{LPF}}(n)$,

$$S_{\varphi}(f) = S_{\varphi_s}(f) \cdot |H_{\text{LPF}}(f)|^2.$$

The phase noise bandwidth B_{φ} is given by 3 dB bandwidth of the filter. The property of phase noise is therefore modeled by the above equation. Generally speaking, there are two kinds of effects from phase noise to the quality of OFDM signals: common phase error and ICI. The second impact is unique for multi-carrier communications such as OFDM; we thus expect that phase noise critically affects OFDM communications, especially for high-order modulation such as 64-QAM.

In addition to common RF factors of radio communication systems, there are some special factors in OFDM systems. The very unique feature is *peak-to-average-power ratio* (PAPR), due to its multi-carrier transmission property. For bandwidth efficient communication, OFDM usually has a large number of sub-carriers and each employs a high-order QAM. There exist some slight chances that most (or even all) sub-carriers have modulation points onto outer constellation points to result in huge peak power (much higher than average power). Such a phenomenon is known as PAPR, which can cause RF operating point into an undesirable region and affects system performance. The most common solution is to use *clipping* circuits once waveform is received. Although clipping is usually good for practical system operations with tolerable performance degradation, we can still use methods for better amplitude recovery, such as decision aided, Bayesian inference, and coding.

9.3   Channel Estiamtion and Equalization

To design channel estimation, we first model the fading channel as

$$h(t, \tau) = \sum_k \alpha_k(t)\delta(\tau - \tau_k),$$

where τ_k is the delay of the kth path and $\alpha_k(t)$ represents the complex amplitude WSS Gaussian process with

$$r_{\alpha_k}(\Delta t) = E\{\alpha_k(t + \Delta t)\alpha_k^*(t)\} = \sigma_k^2 v_t(\Delta t),$$

where σ_k^2 is the average power of the kth path.

The frequency response of time-varying channel at time t is given by

$$H(t, f) = \int_{-\infty}^{\infty} h(t, \tau)e^{-j2\pi f\tau}d\tau = \sum_k \alpha_k(t)e^{k^{-j2\pi f\tau}}.$$

Therefore, the correlation function of the frequency response is given by

$$
\begin{aligned}
r_H(\Delta t, \Delta f) &= E\{H(t + \Delta t, f + \Delta f)H * (t, f)\} \\
&= \sum_k r_{\alpha_k}(\Delta t)e^{-j2\pi\Delta f\tau_k} \\
&= v_t(\Delta t)\sum_k \sigma_k^2 e^{-j2\pi\Delta f\tau_k} \\
&= \sigma_H^2 r_t(\Delta t)r_f(\Delta f),
\end{aligned}
$$

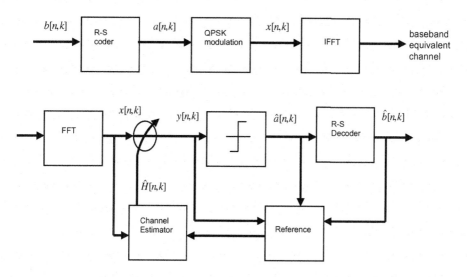

Fig. 9.8 Baseband equivalent example of OFDM transmitter and receiver.

where σ_H^2 is the total average power of the channel impulse response and is given by

$$\sigma_H^2 = \sum_k \sigma_k^2$$

$$r_f(\Delta f) = \sum_k \frac{\sigma_k^2}{\sigma_H^2} e^{-j2\pi\Delta f\tau_k}.$$

Signal at the kth tone and nth block can be represented as

$$x[n,k] = H[n,k]a[n,k] + w[n,k],$$

where $w[n,k]$ is the additive Gaussian noise with zero mean and variance ρ^2. $a[n,k]$ can be estimated as $y[n,k]$ by minimum mean-squared error (MMSE).

$$y[n,k] = \frac{H*[n,k]x[n,k]}{|H[n,k]|^2}.$$

A temporal estimation of $H[n,k]$ is given by

$$\tilde{H}[n,k] = x[n,k]a*[n,k] = H[n,k] + w[n,k]a*[n,k].$$

An MMSE channel estimator can be constructed as

$$\hat{H}[n,k] = \sum_{m=-\infty}^{0} \sum_{l=-(K-k)}^{k-1} c[m,l,k]\tilde{H}[n-m,k-l].$$

Where $c[m,l,k]$ are selected to minimize $E|\hat{H}[n,k] - H[n,k]|^2$ and totally k tones in OFDM.

The above derivation can be based on different criteria to obtain different methods to achieve channel estimation. Channel estimation is a critical step in modern OFDM and even more critical in multiuser OFDM systems. It is still an active research area toward broadband wireless communications using OFDM.

10

Spread Spectrum Communication and Code Division Multiple Access

Spread spectrum (SS) communications have been widely applied in military communication systems for a long time and have been applied in commercial systems since the late 1980s. Qualcomm Inc. founded by I. Jacobs and A. Vertibi proposed to adopt a multiuser version of SS communications known as code division multiple access (CDMA, as a counterpart of frequency division multiple access (FDMA) or time division multiple access (TDMA)) for digital cellular communications. This technology was later standardized as the IS-95 CDMA cellular system, as a version of the so-called second-generation wireless communication systems (i.e., digital cellular) with other narrowband digital systems (such as GSM), while the first-generation wireless communication systems are analog cellular using FDMA. CDMA at wider bandwidth has been adopted in the third-generation wireless communication systems (also known as IMT-2000, International Mobile Telecommunication-2000), and people usually name such a technology as wideband CDMA (WCDMA), including both frequency division multiplexing versions of IMT-2000 in 3GPP and 3GPP2.

The history of SS communications can be found in a paper by R. Sholtz originally published in the special issue of the *IEEE Transactions on Communications*, May 1982, a special issue on SS communications. The origin of SS can be traced back to 1920s at the MIT Lincoln Laboratory. However, it has never been close to a mature and widely applied technology until 1950s and later. The primary purpose to develop an SS technology is for military applications. A good military communication system has to survive in jamming environment and with low probability of interception. SS communication, which was known as "communication by noise", is thus attractive in military communications. Obviously, traditional narrowband digital communication systems are far from to achieve these purposes. A *wideband communication* approach has thus been developed. Note that all systems in this book up to this point are single-carrier or multi-carrier narrowband communication systems.

10.1 Baseband Equivalent SS Communications

We use the system shown in Figure 10.1 to illustrate how an SS works. There are K transmitter–receiver pairs that intend to achieve independent communication simultaneously at the same channel.

Traditional SS systems intend to reach the following functions, especially for military applications.

- Jamming: The desired communication wants to survive under the jammers' interference (Figure 10.2). The way to

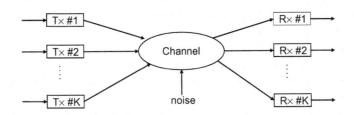

Fig. 10.1 Spread spectrum communications.

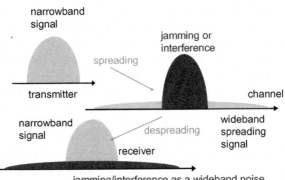

Fig. 10.2 Simple illustration of the spread spectrum communication concept.

achieve this goal is to apply a spread waveform $c_i(t)$ to the ith user's signal so that $r_i(t)$ is a wideband waveform while the jamming interference remains a narrowband waveform while the jamming waveform/interference can be spread to a wideband noise-like waveform. We can use a simplified example to see its feasibility. Suppose that the channel noise is at the level of 1 (unit), and the narrowband signal strength is 100 (i.e., signal-to-noise (SNR) ratio at 20 dB). If we can spread signal by a factor of 10 000 (i.e., 40 dB), the "enemy" observes the channel from noise level 1 to new signal-noise level of 1.01, which is practically not distinguishable. Similarly, the jamming at the receiver after de-spreading is also not distinguishable.

- Low probability of detection/interception: After the $C_i(t)$ spreading feature, waveform $r_i(t)$ spreads its energy form a narrowband to a wideband. This ratio in terms of bandwidth is well known as processing gain, which can be with the order of 104. This kind of ratio makes the power spectral density of $r_i(t)$ at very low level. Consequently, detection or interception becomes extremely difficult and SS communication is the best way for secure communication.

- Ranging: SS signal/modulation usually provides fine phase/timing resolution and thus is good at ranging system, e.g., global positioning system/satellite. However, there are some more attractive features for SS communication in commercial applications.

- Low interference to other systems: Owing to its low psd, SS systems are good at reducing interference to other existing systems. This is an important feature for commercial/civilized applications to deal with crowded spectrum utilization. In the united states, FCC allocating ISM (industrial, scientific, medical) bands with SS requirements illustrates this feature.

- Anti-interference: Similar to anti-jamming capability, SS inherents the capability of anti-interference.

- Multiple-access and system capacity: The present research showed that CDMA can provide higher capability of users in cellular networks and no frequency plan is needed, which is the main reason to result in IS-95 and IMT-2000 cellular systems.

- Anti-multipath: The present personal communication services typically operate in strong multipath fading channels, which cause serious troubles in system design and planning. SS has another inherent feature of removing effects of multipath fading via its built-in time diversity or frequency diversity.

10.1.1   Baseband Equivalent Model

For all of the above purposes, considering baseband equivalent waveforms only, we have to pick up an appropriate waveform $\{C_i(t)\}_{i=1}^{K}$ with the desired features as follows.

It had better been a noise-like waveform, which means the following mathematical properties:

$$\int_0^T C_i(t) = 0 \tag{10.1}$$

$$\int_0^T C_i(t)C_i(t-\tau)dt = N\delta(\tau), \tag{10.2}$$

where T is the symbol/bit period.

$\{C_i(t)\}_{i=1}^{K}$ form a set of orthogonal basis. That is,

$$\int_0^T C_i(t)C_j(t)dt = N\delta_{ij}. \tag{10.3}$$

With finite bandwidth for asynchronous communication, we cannot find $\{C_i(t)\}$ to satisfy (10.1)–(10.3). However, if we loosen the quality to approximating zero, then $\{C_i(t)\}$ is possible to exist. To consider the general behavior of SS communication we temperately do not specify $\{C_i(t)\}$ to look into SS features. Assume that each receiver has a correlation receiver structure, which is optimal for AWGN channels under perfect synchronization; we prove important features of SS communication in the following simplified derivations.

10.1.2 Anti-Interference in Multi-Access

Suppose that the ith user experience a narrrowband interference with the simplest form A_o, that is,

$$r_i'(t) = \sum_{k=1}^{K} d_i(t)C_i(t) + A_0. \tag{10.4}$$

The correlation receiver output for the ith user is given by

$$\hat{d}_i(t) = \int_0^T r_i'(t)C_i(t)dt$$

$$= \int_0^T A_0(t)C_i(t)dt + \int_0^T \left[\sum_{k=1}^{K} d_i(t)C_k(t)\right] C_i(t)dt$$

$$= \int_0^T A_0 C_i(t)dt + \sum_{k=1}^{K}\int_0^T d_i C_k(t)C_i(t)dt. \tag{10.5}$$

Looking at the first term, by (10.1) we have

$$A_0 \int_0^T dC_i \approx A_{0i} \cdot 0 \approx 0 \quad 0 \le t \le T.$$

Looking at the second term, by (10.3) we have

$$\sum_{k=1}^{K}\int_0^T d_k C_k(t)C_i(t)dt \approx N d_i \quad 0 \le t \le T.$$

Note that $d_i(t) = d_i$, a constant, over $(0,T)$. Consequently, we have

$$\hat{d}_i(t) = N d_i \quad 0 \le t \le T.$$

We successfully demodulate d_i in a multi-access channel even with narrowband interference.

10.1.3 Anti-multipath

Suppose that the channel can be modeled as a linear filter with impulse response $h(t)$. This model is very similar to the well-accepted two-ray multipath model in urban environment, that is,

$$h(t) = \delta(t) + \alpha\delta(t - \tau), \tag{10.6}$$

where $0 < \alpha < 1$, $0 < \tau < T$ (in most cases, $\tau \ll T$)

$$r_i'(t) = r_i(t) \times h(t)$$
$$= d_i C_i(t) + \alpha d_i C_i(t - \tau) + \alpha d_{i-1} C_i(t + T - \tau). \quad (10.7)$$

The third term is due to the memory of the channel, that is,

$$\hat{d}_i(t) = \int_i^t r_i'(t) C_i(t) dt$$

$$= d_i \int_i^t C_i(t) C_i(t) dt + \alpha d_i \int_i^t C_i(t - \tau_\alpha) C_i(t) dt$$

$$+ \alpha d_{i-1} \int_i^t C_i(t + T - \tau) C_i(t)$$

$$\approx d_i \cdot N. \quad (10.8)$$

The SS system can thus overcome the multipath effects from the channels. Although this is a very simplified explanation, it is very intuitive with certain mathematical precision.

There are two fundamental ways to realize the $c(t)$ stated above. The first approach is known as direct sequence SS (DS-SS), and the second approach is to consider the time average of hopping narrowband signals known as frequency-hopping (FH) SS communications.

10.2   DS-SS Communications

The idea of DS-SS communications is to find a code sequence with desired features whose waveform is used as $c(t)$. Figure 10.3 illustrates an example to spread the bits by using a code sequence in time domain. The spreading code sequence consists of higher-rate codes. Each bit in the spreading code sequence is called a *chip*. In DS-SS systems, the resulting spread waveform commonly has much higher rate than the original information bit/symbol rate. The ratio of chip rate to symbol rate is known as *processing gain*.

Unfortunately, such a code does not perfectly exist. What we can do most is to find codes with features close to the ideal case. Several codes have been known to us until now. The most common one may

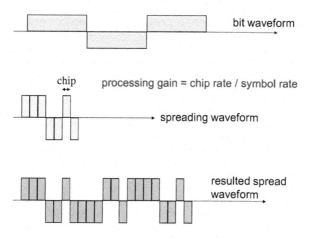

Fig. 10.3 Spreading waveform using a code sequence.

be the *maximum length code* (or *m-sequence*) which has the following correlation properties: Introducing a function $\xi\;(\cdot)$ such that $\xi(\alpha) = (-1)^{\alpha+1}$, where $\alpha \in \{0,1\}$, we map an m-sequence u into a binary polarity sequence $\xi(u)$.

(a) The period of u and $\xi(u)$ is $L = 2^M - 1$, where L is the length of the m-sequence that can be generated by M shift registers.

(b) There are exactly L non-zero sequences generated by one polynomial, which is the multiplication of some primitive polynomials in $GF\ (2^M)$. Furthermore, they are $u, Tu, \ldots, T^{L-1}\ u$, where T is the operator of left circularly shift by one digit. They can also be considered as code phases of u. $\xi(u)$ also has the same property.

(c) (Shift and Add Property) Given distinct integers $i, j, 0 \le L$, $j \le L-1$, there exists a unique integer k that is not equal to i and j, such that $0 \le k \le L-1$ and $T^i u \oplus T^j u = T^k u$, where $\oplus$ represents addition with mod 2 operation.

(d) There are $2^{M-1} = (L+1)/2$ 1's and $2^{M-1} - 1 = (L-1)/2$ 0's in an m-sequence. We define the number of 1's in u as the weight of u, $W(u)$ and the corresponding definition of the weight of $\xi(u), W_\xi(u)$.

(e) (Autocorrelation Property) The periodic autocorrelation function is defined as

$$R_{\xi(u)}(l) = \sum_{i=1}^{L} \xi(u_i)\xi(u_{i+l})$$

$$= \sum_{i=1}^{L} (-1)^{u_i \oplus u_{i+l}}$$

$$= \sum_{i=1}^{L} \xi(u_i \oplus u_{i+l}), \qquad (10.9)$$

where u_i is the ith bit/digit of the m-sequence. Associating with property (d), the m-sequence has a two-valued autocorrelation function for the NRZ signals

$$R_{\xi(u)}(l) = \begin{cases} N, & l \equiv 0 \mod L \\ -1, & l \not\equiv 0 \mod L. \end{cases} \qquad (10.10)$$

Figure 10.4 shows an example of a maximum length code with generating polynomial $1 + X + X^4$, and its autocorrelation calculated by circular shifting the code sequence.

We can also verify all properties through this simple length-7 m-sequence. It is the longest code sequence that can be generated

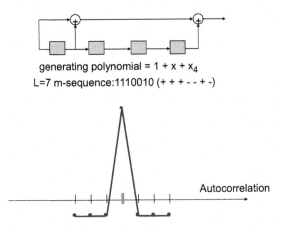

generating polynomial = 1 + x + x$_4$
L=7 m-sequence:1110010 (+ + + - - + -)

Autocorrelation

Fig. 10.4 Maximum length code and its autocorrelation function.

Table 10.1 Run-length distribution of a 127-chip maximum length code.

Run length	No. of 1 runs	No. of 0 runs	No. of chips
1	16	16	32
2	8	8	32
3	4	4	24
4	2	2	16
5	1	1	10
6	0	1	6
7	1	0	7
Total			127

by a given number of stages of delays. This is why such codes are called maximal length codes. Of course, the generating polynomial is determined from cyclic code theory. We can further verify the nearly perfect randomness of maximum length codes as the following table for a 127 m-sequence by checking run-length distribution.

We apply such a kind of codes to spread the signal waveform into a wideband signal. The ratio of waveform spreading is defined as the *processing gain*, *PG*. That is,

$$PG = \frac{B_{\text{afters spreading}}}{B_{\text{before spreading}}}. \tag{10.11}$$

Processing gain is a key factor to define DS-SS system. Equation (10.11) is defined based on the expansion factor of bandwidth, which is the reason why we call such a system as "SS" system. There are several ways to define the bandwidth, while the common way is to define through 3-dB bandwidth or 6-dB bandwidth (for the power). Note the difference between this definition and Figure 10.3. We usually use the definition in Figure 10.3 for the theory of coherent DS-SS, but industrial regulation and non-coherent DS-SS widely use (10.11).

A typical DS-SS communication system has the following block diagram as shown in Figure 10.5. There exist two mathematically equivalent realizations while they are different in terms of circuit implementation. We may multiply the spreading waveform (i.e., m-sequence in most situations) with the symbol sequence prior to modulator or after modulator. This is typically known as *coherent* DS-SS, which is widely used in commercial systems such as IS-95 CDMA cellular, IMT-2000,

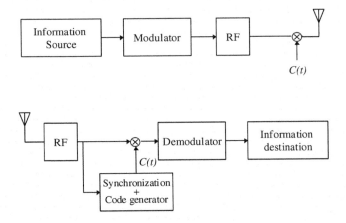

Fig. 10.5 Block diagram of a DS-SS communication system.

and IEEE 802.11b wireless LANs. The way as shown in Figure 10.5 is known as *non-coherent* DS-SS, which is a typical system structure and has been applied in many military communication systems. At the receiver end, a primary difference with traditional narrowband receiver lies in the synchronization of DS-SS in order to correlate spreading waveform to recover the original narrowband signal waveform. Therefore, the DS-SS receiver generally consists of several subsystems in addition to RF/IF circuits. They are

- an SS synchronizer to derive the timing of spreading waveform;
- a carrier recovery scheme to cancel the effects of carrier and Doppler spread;
- a baseband processor for demodulation, timing recovery (especially if the spreading waveform does not have the same period as the symbol period), and other possible signal processing functions such as equalization, forward error-correcting code decoding, etc.

The immediate challenge to design a non-coherent (and also a coherent) DS-SS system is the low *SNR* at the receiver front-end to result in difficulty to recover parameters for signal detection, especially synchronization so that the receiver can re-generate spreading codes at correct timing for demodulation. The common synchronization techniques can be applied for *SNR* of at least 5–10 dB, while the common DS-SS systems work at low *SNR* and even negative *SNR* for most cases. Typical approach to resolve this challenge is to adopt two-stage synchronization (rare in narrowband communication systems):

- Acquisition (coarse synchronization): *Acquisition* is a technique using integrate-and-dump circuits such as correlation receiver to detect the appearance of a signal based on a presumed timing. The dwell time is usually multiple symbol periods to enhance effective detection. The number of timing candidates is less than exactly needed to speed up finding rough timing. The timing resolution is determined from the number of timing candidates that are usually equal to the number of chips per symbol or its multiple. Since acquisition targets at identifying rough timing, it is also called coarse synchronization. Figure 10.6 depicts a single-dwell serial search code acquisition scheme for DS-SS, whose behaviors can be analyzed by a Markov chain model. To further speed up acquisition, we may use multiple-dwell schemes with parallel search (timing candidates).

- Tracking (fine synchronization): After initial acquisition, we adopt a code *tracking* mechanism to lock the fine timing,

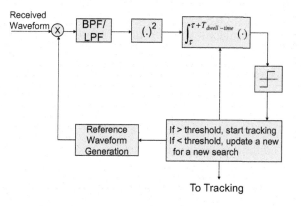

Fig. 10.6 Single-dwell code acquisition.

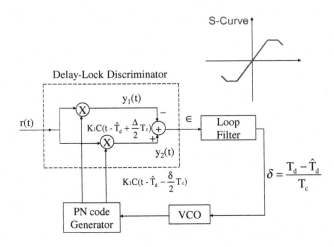

Fig. 10.7 DLL.

which is also known as fine synchronization. Coding tracking is usually realized by a closed loop and delay-locked loop (DLL, as illustrated in Figure 10.7) based on the early-late-gate synchronizer, which is the most well-known scheme. The delay-locked discriminator is obtained by the same concept as the S-curve in the early-late-gate synchronizer, by further taking pseudo-noise (PN) code correlation into consideration.

Obviously, the goal of code acquisition is to obtain the rough estimate of timing by detection theory and effective search algorithms. A number of its variations that we just introduced (single-dwell serial-search) can be developed based on the technology such as multi-dwell correlations, parallel search, and sequential detection. In fact, for any signal transmission with low *SNR*, the spirit of code acquisition can be applied.

The purpose of PN code tracking is to track signal timing and to lock it, once code acquisition is done. To avoid potential false acquisition, confirmation test of signal acquisition is usually required prior to code tracking. In the early days of SS communications, two correlation branches in DLL are not favored all the time, due to the increasing implementation complexity. A well-known variation of DLL

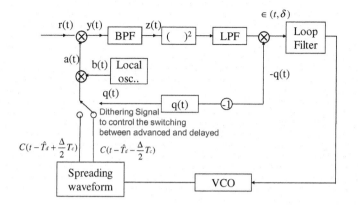

Fig. 10.8 TDL for non-coherent DS-SS.

using only one correlation branch is called tau-dither loop (TDL), which applies a dithering signal by equally partitioning observation interval into two halves for advance and delay versions of correlation as shown in Figure 10.8. The immediate advantage from TDL is the saving of complexity at the price of losing around $3\,\text{dB}$ *SNR*, due to half observation period compared with DLL.

10.3 FH-SS Systems

As any scientific observation is based on *time average* one way or another, another realization of SS communication systems is the so-called FH system which hops its narrowband signal over a wide range of frequency band by controlling IF frequency synthesizer (Figure 10.9). As long as such hopping is fast enough (i.e., much faster than enemy's observation frequency or equivalently much shorter than enemy's observation interval), enemy's observation is just the time average of the spectrum and thus difficult to observe any communication.

To facilitate FH-SS, we require a high-quality frequency synthesizer to hop narrowband signals/waveforms at extremely high speed, typically thousands of hops, under a designate PN hopping code. It is never easy to implement such a frequency synthesizer. Modern frequency synthesizers are typically digital through a special-purpose high-speed calculation circuits. The PC hopping code is specially selected to minimize

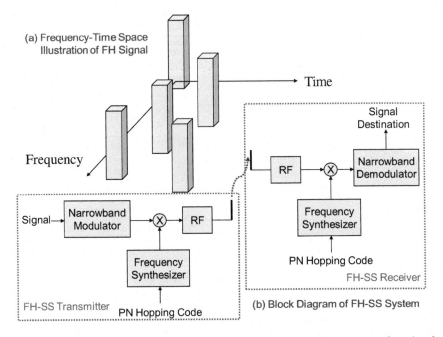

(a) Frequency-Time Space
    Illustration of FH Signal

Fig. 10.9 FH-SS Communications: (a) frequency-time domain illustration of a signal;
(b) block diagram of a transmitter and a receiver.

cross-correlation among coexisting FH-SS signal transmissions, which
is just like PN codes in DS-SS systems and that is why we call them
PN hopping codes. However, due to different code natures, PN hopping codes are usually different from the m-sequences in DS-SS, as the
purpose of PN hopping codes is to minimize the number of "hits" (collisions from simultaneously multiple transmission at the same frequency)
and other interference concerns.

FH-SS systems are widely applied in military communications due
to their nature that is difficult to jam the signals. There are a number of FH jamming strategies and anti-jamming strategies in literatures. Interesting readers may find details in a well-documented book
authored by R. Scholtz, J. Omura et al. The most widely applied FH-
SS system might be *Bluetooth*, which hops 1600 times per second over
2.4 GHz ISM band with 75 pre-selected hopping sequences to support
up to 1 Mbps signal of 1 MHz bandwidth using Gaussian-filtered FSK
modulation.

10.4 CDMA and Multiuser Communications

The major commercial applications of DS-SS system are due to its multiple access capability, which is known as CDMA, as a counterpart of FDMA and TDMA. CDMA has been first proposed and used in IS-95 cellular at a chip rate of 1.2288 Mcps and then WCDMA at 5 MHz bandwidth or its multiples became the air-interface technology for IMT-2000 (also widely referred to as third-generation wireless communications). In this section, we briefly introduce the primary technology to implement CDMA cellular.

10.4.1 Spreading Codes

To achieve multiple access purpose, the spreading codes (or signature sequences) must have good cross-correlation property, in addition to autocorrelation property for a single link (i.e., transmitter–receiver pair). m-sequences have nice autocorrelation but not cross-correlation to achieve multiple access due to the shift-and-add property mentioned in Section 10.2. We have to seek another class of sequences for such a purpose. The most popular class is *Gold codes*, which are derived from the m-sequences, by adding two pre-selected and independently generated m-sequences together. Gold summarized the generator polynomial pairs to enforce the side-lobes of cross-correlation, $\varphi(c_i, c_j)$, bounded by

$$\varphi(c_i, c_j) \leq \begin{cases} 2^{(n+1)/2} + 1, & n \text{ odd} \\ 2^{(n+2)/2} - 1, & n \text{ even.} \end{cases} \tag{10.12}$$

Figure 10.10 depicts this scenario. In fact, Gold codes have been widely used in asynchronous CDMA, such as Tracking and Data Relay Satellite System (TDRSS) for US military and aerospace communications, and GPS for positioning.

For CDMA cellular systems that are synchronous, Walsh–Hadamard sequences (or Walsh codes) are adopted to ensure orthogonal among users. Figure 10.10(b) depicts the generation of Walsh codes with length 2^n, from the basic matrix by duplicating and weighting by -1 following the basic matrix.

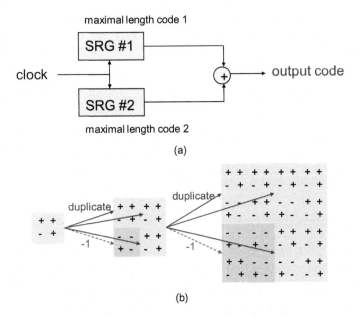

Fig. 10.10 (a) Gold codes generation; (b) Walsh codes generation for code length 2, 4, 8.

10.4.2   Near–Far Problem

The concept for CDMA is nice but there exists a fundamental challenge, *near–far* problem, slightly similar to capture effect in FM. As CDMA allows simultaneous transmissions over the same frequency band, the signal from the far location (which suggests low signal strength) arrives a specific receiver with much lower power level than that from the near location (which suggests high signal strength). In principle, both signals should be received well. However, due to the RF front-end design, the small signal might be lost in practice. One way of resolving this challenge is to use *power control*, which lowers the signal transmission power for near locations and increases the signal transmission power for far locations, so that the signal strength at the receiver (that is, base station in cellular systems) from different locations would be around the same. WCDMA system updates power control up to 1600 times per second.

10.4.3   Multiuser Detection (MUD)

With applications of CDMA, one receiver (e.g., base station) may simultaneously receive multiple signals from multiple transmitters, or a system may consist of multiple transmitter–receiver pairs at the same frequency band and at the same time. In the past, such a situation results in collision of signals to lose all signals or interference from other users known as multiple access interference (MAI). However, a revolutional view by S. Vender and V. Poor is to treat such signals from different transmitters as a vector of signal and the receiver demodulates this signal vector to obtain the estimates for all signals (signal components in a signal vector), which provides the optimal detection and MAI cancellation. It was also shown that MUD provides near–far resistance in detection. All these nice features suffer non-polynomial complexity for the receiver. Of course, various sub-optimal receiver structures have been proposed in literatures. Last but not least, MUD can be applied for general multiuser communications, not limited only to CDMA.

References

[1] M. Abramowitz and I. A. Stegun, *Handbook of Mathematical Functions with Formulas, Graphs, and Mathematical Tables*. New York: Dover Publications, 1965.

[2] N. Abramson, *Information Theory and Coding*. New York: McGraw-Hill, 1963.

[3] Y. Akaiwa, *Introduction to Digital Mobile Communication*. New York: Wiley, 1997.

[4] J. B. Anderson, T. Aulin, and C. E. Sundberg, *Digital Phase Modulation*. New York: Plenum Publishers, 1986.

[5] J. B. Anderson and S. Mohan, *Source and Channel Coding; An Algorithmic Approach*. Boston, Mass: Kluwer Academic, 1991.

[6] J. B. Anderson, *Digital Transmission Engineering*. Piscataway, NJ: IEEE Press, 1999.

[7] R. B. Ash, *Information Theory*. New York: Wiley, 1965.

[8] Bell Laboratories Technical Staff, *A History of Engineering Science in the Bell System: The Early Years (1875–1925)*, Books on Demand, Ann Arbor, Michigan: 1975.

[9] J. C. Bellamy, *Digital Telephony*. New York: Wiley, second ed., 1991.

[10] S. Benedetto and E. Biglieri, *Principles of Digital Transmission with Wireless Applications*. New York: Kluwer Academic/Plenum Publishers, 1999.

[11] S. Benedetto, E. Biglieri, and V. Castellani, *Digital Transmission Theory*. Englewood Cliffs, NJ: Prentice-Hall, 1987.

[12] W. R. Bennett, *Introduction to Signal Transmission*. New York: McGraw-Hill, 1970.

[13] K. B. Benson and J. C. Whitaker, *Television Engineering Handbook*. New York: McGraw-Hill, rev. ed., 1992.

[14] T. Berger, *Rate Distortion Theory: A Mathematical Basis for Data Compression*. Englewood Cliffs, NJ: Prentice-Hall, 1971.

[15] E. R. Berlekamp (ed.), *Key Papers in the Development of Coding Theory*. Piscataway, NJ: IEEE Press, 1974.

[16] D. Bertsekas and R.Gallager, *Data Networks*. Prentice-Hall, second ed., 1992.

[17] V. K. Bhargava, D. Haccoun, R. Matyas, and P. Nuspl, *Digital Communications by Satellite: Modulation, Multiple Access, and Coding*. New York: Wiley, 1981.

[18] E. Biglieri, D. Divsalar, P. J. McLane, and M. K. Simon, *Introduction to Trellis-Coded Modulation with Applications*. New York: Macmillan, 1991.

[19] J. A. C. Bingham, *The Theory and Practice of Modem Design*. New York: Wiley, 1988.

[20] R. B. Blachman and J. W. Tukey, *The Measurement of Power Spectra, from the Point of View of Communication Engineering*. New York: Dover, 1958.

[21] R. E. Blahut, *Principles and Practice of Information Theory*. Reading, Mass: Addison-Wesley, 1987.

[22] R. E. Blahut, *Digital Transmission of information*. Reading, Mass: Addison-Wesley, 1990.

[23] G. E. P. Box and G. M. Jenkins, *Time Series Analysis: Forecasting and Control*. San Francisco: Holden Day, 1976.

[24] R. N. Bracewell, *The Fourier Transform and Its Applications*. New York: McGraw-Hill, second ed., rev. 1986.

[25] L. Brillouin, *Science and Information Theory*. New York: Academic Press, second ed., 1962.

[26] K. W. Cattermole, *Principles of Pulse-code Modulation*. New York: American Elsevier, 1969.

[27] W. Y. Chen, DSL: Simulation techniques and standards development for digital subscriber line systems, Indianapolis, Ind: Macmillan Technical Publishing, 1998.

[28] K. C. Chen and R. DeMarca (ed.), *Mobile WiMAX*. Wiley-IEEE, 2008.

[29] T. D. Chiueh and P. Y. Tsai, OFDM Based Receiver Design for Wireless Communications, 2007.

[30] J. M. Cioffi, *Digital Data Transmission*. EE379C Course Textbook, Stanford University, 1998.

[31] G. C. Clark, Jr., and J. B. Cain, *Error-correction Coding for Digital Communications*. New York: Plenum Publishers, 1981.

[32] Communication Society, *History of Communications*. IEEE, 2002.

[33] T. M. Cover and J. A. Thomas, *Elements of Information Theory*. New York: Wiley, 1991.

[34] H. Cramér and M. R. Leadbetter, *Stationary and Related Stochastic Processes: Sample Function Properties and Their Applications*. New York: Wiley, 1967.

[35] W. B. Davenport. Jr., and W. I. Root, *An Introduction to the Theory of Random Signals and Noise*. New York: McGraw-Hill, 1958.

[36] R. C. Dixon, *Spread Spectrum Systems*. New York: Wiley, second ed., 1984.

[37] R. C. Dixon (ed.), *Spread Spectrum Techniques*. New York, IEEE Press, 1976.

[38] L. J. Doob, *Stochastic Processes*. New York: Wiley, 1953.

[39] J. J. Downing, *Modulation Systems and Noise*. Englewood Cliffs, NJ: Prentice-Hall, 1964.

[40] W. Feller, *An Introduction to Probability Theory and Its Application*. New York: Wiley, vol. 1, third ed., 1968.

[41] T. L. Fine, Theories of Probability: *An Examination of Foundations*. New York: Academic Press, 1973.

[42] L. E. Franks (ed.), *Data Communication: Fundamentals of Baseband Transmission*. Dowden, Hutchison, and Ross, 1974.

[43] L. E. Franks, *Signal Theory*. Englewood Cliffs, NJ: Prentice-Hall, 1969.

[44] R. L. Freeman, *Telecommunications Transmission Handbook.* New York: Wiley, fourth ed., 1998.

[45] B. J. Frey, *Graphical Models for Machine Learning and Digital Communications.* Cambridge, Mass: MIT Press, 1998.

[46] R. G. Gallager, *Information Theory and Reliable Communication.* New York: Wiley, 1968.

[47] R. G. Gallager, *Low-Density Parity-Check Codes.* Cambridge, Mass: MIT Press, 1963.

[48] F. M. Gardner, *Phaselock Techniques.* New York: Wiley, second ed., 1979.

[49] V. K. Gary and J. E. Wilkes, *Principles & Applications of GSM.* Englewood Cliffs, NJ: Prentice-Hall, 1999.

[50] A. Gersho and R. M. Gray, *Vector Quantization and Signal Compression.* Boston, Mass: Kluwer Academic, 1992.

[51] J. D. Gibson (ed.), *The Mobile Communications Handbook.* Piscataway, NJ: IEEE Press, 1996.

[52] R. D. Gitlin, J. F. Hayes, and S. B. Weinstein, *Data Communications Principles.* New York: Plenum, 1992.

[53] B. Goldberg and H. S. Bennett (ed.), *Communication Channels: Characterization and Behavior.* New York: IEEE Press, 1976.

[54] S. W. Golomb (ed.), *Digital Communications with Space Applications.* Englewood Cliffs, NJ: Prentice-Hall, 1964.

[55] S. W. Golomb, *Shift Register Sequences.* San Francisco: Holden-Day, 1967.

[56] Gordon Stuber, Principles of Mobile Communications, 1996.

[57] R. M. Gray and L. D. Davisson, *Random Processes: A Mathematical Approach for Engineers.* Englewood Cliffs, NJ: Prentice-Hall, 1986.

[58] R. M. Gray and L. D. Davisson, *Random Processes.* Prentice-Hall, 1986.

[59] P. E. Green, Jr., *Computer Network Architectures and Protocols.* New York: Plenum, 1982.

[60] P. E. Green, Jr., *Fiber Optic Networks.* Englewood Cliffs, NJ: Prentice-Hall, 1993.

[61] M. S. Gupta (ed.), *Electrical Noise: Fundamentals and Sources.* New York: IEEE Press, 1977.

[62] R. W. Hamming, *Coding and Information Theory.* Englewood Cliffs, NJ: Prentice-Hall, 1980.

[63] L. Hanzo, W. Webb, T. Keller, *Single- and Multi-carrier Quadrature Amplitude Modulation.* Wiley, 2000.

[64] S. Haykin, *Adaptive Filter Theory.* Englewood Cliffs, NJ: Prentice-Hall, third ed., 1996.

[65] S. Haykin and B. Van Veen, *Signals and Systems.* New York: Wiley, 1999.

[66] S. Haykins, *Communication Systems.* John Wiely & Sons, fourth ed., 2001.

[67] C. Heegard and S. B. Wicker, *Turbo Coding.* Boston, Mass: Kluwer Academic Publishers, 1999.

[68] G. Held, *The Complete Modem Reference.* New York: Wiley, third ed., 1997.

[69] C. W. Helstrom, *Statistical Theory of Signal Detection.* Elmsford, NY: Pergarnon Press, 1968.

[70] C. W. Helstrom, *Probability and Stochastic Processes for Engineers*. New York: Macmillan, second ed., 1990.

[71] J. K. Holmes, *Coherent Spread Spectrum Systems*. New York: Wiley, 1982.

[72] W. C. Jakes, Jr., (ed.), *Microwave Mobile Communications*. New York: Wiley, 1974.

[73] N. S. Jayant and P. Noll, *Digital Coding of Waveforms: Principles and Applications to Speech and Video*. Englewood Cliffs, NJ: Prentice-Hall, 1984.

[74] N. S. Jayant (ed.), *Waveform Quantization and Coding*. New York: IEEE Press, 1976.

[75] M. C. Jeruchim, B. Balaban, and J. S. Shanmugan, *Simulation of Communication Systems*. New York: Plenum, 1992.

[76] E. C. Jordan and K. G. Balmain, *Electromagnetic Waves and Radiating Systems*. Englewood Cliffs, NJ: Prentice Hall, second ed., 1968.

[77] S. Karp, R. M. Gagliadi, S. E. Moran, and L. B. Stotts, *Optical Channels*. Plenum Press, 1988.

[78] A. Khintchin, *Mathematical Foundations of Information Theory*. New York: Dover, 1957.

[79] A. N. Kolmogorov, *Foundations of the Theory of Probability*. New York: Chelsea Publishing, 1956.

[80] J. Laiho, A. Wacker, and T. Novosad, (ed.), *Radio Network Planning and Optimization for UMTS*. Wiley, 2006.

[81] B. P. Lathi, *Modern Digital and Analog Communication Systems*. Oxford University Press, second ed., 1995.

[82] I. Lebow, *Information Highways and Byways*. Piscataway, NJ: IEEE Press, 1995.

[83] E. A. Lee and D. G. Messerschmitt, *Digital Communication*. Boston, Mass: Kluwer Academic, second ed., 1994.

[84] J. S. Lee and L. E. Miller, *CDMA Systems Engineering Handbook*. Boston, Mass: Artech House Publishers, 1998.

[85] Y. W. Lee, *Statistical Theory of Communication*. New York: Wiley, 1960.

[86] W. C. (Y.) Lee, *Mobile Communications Engineering*. New York: McGraw-Hill, 1982.

[87] A. Leon-Garcia, *Probability and Random Processes for Electrical Engineering*. Reading, Mass: Addison-Wesley, second ed., 1994.

[88] Y. Li, G. Stuber (ed.), *Orthogonal Frequency Division Multiplexing for Wireless Communications*. Springer, 2006.

[89] S. Lin and D. J. Costello, Jr., *Error Control Coding: Fundamentals and Applications*. Englewood Cliffs, NJ: Prentice-Hall, 1983.

[90] W. C. Lindsey, *Synchronization Systems in Communication and Control*. Englewood Cliffs, NJ: Prentice-Hall, 1972.

[91] W. C. Lindsey and M. K. Simon (ed.), *Phase-locked Loops and Their Applications*. New York: IEEE Press, 1978.

[92] W. C. Lindsey and M. K. Simon, *Telecommunication Systems Engineering*. Englewood Cliffs, NJ: Prentice-Hall, 1973.

[93] M. Loéve, *Probability Theory*. Princeton, NJ: Van Nostrand, 1963.

[94] R. W. Lucky, J. Salz, and E. J. Weldon, Jr., *Principles of Data Communication*. New York: McGraw-Hill, 1968.

[95] F. J. MacWilliams and N. J. A. Sloane, *The Theory of Error-correcting Codes*. Amsterdam: North-Holland, 1977.

[96] V. K. Madisetti and D. B. Williams (ed.), *The Digital Signal Processing Handbook*. Piscataway, NJ: IEEE Press, 1998.

[97] R. J. Marks, *Introduction to Shannon Sampling and Interpolation Theory*. New York/Berlin: Springer-Verlag, 1991.

[98] J. C. McDonald (ed.), *Fundamentals of Digital Switching*. New York: Plenum, second ed., 1990.

[99] R. McDonough and A. D. Whalen, *Detection of Signals in Noise*. New York: Academic Press, second ed., 1995.

[100] R. J. McEliece, *The Theory of Information and Coding: A Mathematical Framework for Communication*. Reading, Mass: Addison-Wesley, 1977.

[101] A. Mengali and N. D'Andrea, *Synchronization Techniques for Digital Receivers*. New York: Plenum, 1997.

[102] H. Meyr and G. Ascheid, *Synchronization in Digital Communications*. New York: Wiley, vol. 1 1990.

[103] H. Meyr, M. Moeneclaey, and S. A. Fechtel, *Digital Communication Receivers: Synchronization, Channel Estimation and Signal Processing*. New York: Wiley, 1998.

[104] A. M. Michelson and A. H. Levesque, *Error-control Techniques for Digital Communication*. New York: Wiley, 1985.

[105] D. Middleton, *An Introduction to Statistical Communication Theory*. New York: McGraw-Hill, 1960.

[106] P. F. Panter, *Modulation, Noise and Spectral Analysis, Applied to Information Transmission*. New York: McGraw-Hill, 1965.

[107] A. Papoulis, *Probability, Random Variables, and Stochastic Processes*. New York: McGraw-Hill, second ed., 1984.

[108] J. D. Parsons, *The Mobile Radio Propagation Channel*. New York: Wiley, 1992.

[109] K. Pahlavan and A. H. Levesque, *Wireless Information Networks*. New York: Wiley, 1996.

[110] J. Pearl, Probabilistic reasoning in intelligent systems: Networks of plausible inference. San Mateo, Calif: Morgan Kaufman Publishers, 1988.

[111] W. W. Peterson and E. J. Weldon, Jr., *Error Correcting Codes*. Cambridge, Mass: MIT Press, second ed., 1972.

[112] J. R. Pierce, *Symbols, Signals and Noise: The Nature and Process of Communication*. New York: Harper, 1961.

[113] H. V. Poor, *An Introduction to Signal Detection and Estimation*. New York/Berlin: Springer-Verlag, second ed., 1994.

[114] T. Pratt and C. W. Bostian, *Satellite Communications*. New York: Wiley, 1986.

[115] J. G. Proakis, C. M. Rader, F. Ling, and C. L. Nikias, *Advanced Digital Signal Processing*. MacMillan, 1992.

[116] J. G. Proakis, *Digital Communications*. New York: McGraw-Hill, third ed., 1995.

[117] J. Proakis and D. Manolakis, *Introduction to Digital Signal Processing.* Macmillan, 1998.

[118] L. R. Rabiner and R. W. Schafer, *Digital Processing of Speech Signals.* Englewood Cliffs, NJ: Prentice-Hall, 1978.

[119] R. Prasad, Universal Wireless Personal Communications, 1998.

[120] K. R. Rao and P. Yip, *Discrete Cosine Transform: Algorithms, Advantages, Applications.* New York: Academic Press, 1990.

[121] T. S. Rappaport, *Smart Antennas.* Piscataway, NJ: IEEE Press, 1998.

[122] T. S. Rappaport, *Wireless Communications: Principles and Practice.* Piscataway, NJ: IEEE Press, 1996.

[123] S. Redl, M. K. Weber, and M. W. Oliphant, *An Introduction to GSM.* Artech House, 1995.

[124] S. O. Rice, "Noise in FM receivers," in *Proceedings of the Symposium on Time Series Analysis*, M. Rosenblatt (ed.), New York: Wiley, pp. 395–411, 1963.

[125] A. Richardson, *WCDMA Design Handbook.* Cambridge, 2005.

[126] J. H. Roberts, Angle Modulation: *The Theory of Systems Assessment.* IEE Communication Series 5 London: Institution of Electrical Engineers, 1977.

[127] S. M. Ross, *Introduction to Probability Models.* Academic Press, 1989.

[128] H. E. Rowe, *Signals and Noise in Communication Systems.* Princeton, NJ: Van Nostrand, 1965.

[129] M. Schwartz, W. R. Bennett, and S. Stein, *Communication Systems and Techniques.* New York: McGraw-Hill, 1966.

[130] M. Schwartz, *Information Transmission, Modulation and Noise: A Unified Approach.* New York: McGraw-Hill, third ed., 1980.

[131] M. Schwartz, *Telecommunication Networks: Protocols, Modeling, and Analysis.* Reading, Mass: Addison Wesley, 1987.

[132] C. E. Shannon and W. Weaver, *The Mathematical Theory of Communication.* Urbana: University of Illinois Press, 1949.

[133] A. N. Shiyayev, *Probability.* Springer-Verlag, 1984.

[134] G. J. Simmons (ed.), *Contemporary Cryptology: The Science of Information Integrity.* Piscataway, NJ: IEEE Press, 1992.

[135] M. K. Simon, J. K. Omura, R. A. Scholtz, and B. K. Levitt, *Spread Spectrum Communications.* New York: Computer Science Press, vols. I, II, and III, 1985.

[136] B. Sklar, *Digital Communications: Fundamentals and Applications.* Englewood Cliffs, NJ: Prentice-Hall, 1988.

[137] D. Slepian (ed.), *Key Papers in the Development of Information Theory.* New York: IEEE Press, 1974.

[138] N. J. A. Sloane and A. D. Wyner, *Claude Shannon: Collected Papers.* Piscataway, NJ: IEEE Press, 1993.

[139] I. S. Sokolnikoff and R. M. Redheffer, *Mathematics of Physics and Modern Engineering.* New York: McGraw-Hill, 1966.

[140] Special Issue on Spread Spectrum Communication. *IEEE Transactions On Communications.* May 1982.

[141] J. J. Spilker, Jr., *Digital Communications by Satellite.* Englewood Cliffs, NJ: Prentice-Hall, 1977.

[142] W. Stallings, *ISDN and Broadband ISDN*. New York: Macmillan, second ed., 1992.

[143] T. Starr, J. M. Cioffi, and P. J. Silverman, *Understanding Digital Subscriber Line Technology*. Englewood Cliffs, NJ: Prentice-Hall 1999.

[144] R. Steele, *Delta Modulation Systems*. New York: Wiley, 1975.

[145] R. Steele and L. Hanzo (eds.), *Mobile Radio Communications*. New York: Wiley, second ed., 1999.

[146] J. J. Stiffler, *Theory of Synchronous Communications*. Englewood Cliffs, NJ: Prentice-Hall, 1971.

[147] A. S. Tanenbaum, *Computer Networks*. Englewood Cliffs, NJ: Prentice-Hall, second ed., 1995.

[148] W. H. Tranter, D. P. Taylor, R. E. Ziemer, N. F. Maxemchuk, and J. W. Mark (ed.), *The Best of The Best: 50 Years of Communications and Networking Research*. IEEE, 2007.

[149] S. Tantaratana and K. M. Ahmed, *Wireless Applications of Spread Spectrum Systems: Selected Readings*. Piscataway, NJ: IEEE Press, 1998.

[150] J. Terry and J. Heiskala, OFDM wireless LANs, Sams Publishing, 2002.

[151] J. Thomas, Introduction to Statistical Communications, 1968.

[152] T. M. Thompson, *From Error-correcting Codes Through Sphere Packings to Simple Groups*. Washington DC: The Mathematical Association of America, 1983.

[153] D. J. Torrieri, *Principles of Military Communication Systems*. Boston, Mass: Artech House Publishers, second ed., 1992.

[154] W. Trappe and L. C. Washington, *Introduction to Cryptography with Coding Theory*. Prentice Hall, 2002.

[155] A. Van der Ziel, *Noise: Source, Characterization, Measurement*. Englewood Cliffs, NJ: Prentice-Hall, 1970.

[156] H. L. Van Trees, *Detection, Estimation, and Modulation Theory*. New York: Wiley, Part I, 1968.

[157] S. Verdú, *Multiuser Detection*. New York, USA: Cambridge University Press, 1998.

[158] A. J. Viterbi, *Principles of Coherent Communication*. New York: McGraw-Hill, 1966.

[159] A. J. Viterbi and J. K. Omura, *Principles of Digital Communication and Coding*. New York: McGraw-Hill, 1979.

[160] A. Viterbi, *CDMA: Principles of Spread Spectrum Communication*. Addison-Wesley, 1995.

[161] G. N. Watson, *A Treatise in the Theory of Bessel Functions*. New York: Cambridge University Press, second ed., 1966.

[162] S. B. Wicker and V. K. Bhargava (eds.), *Reed-Solomon Codes*. Piscataway, NJ: IEEE Press, 1994.

[163] B. Widrow and S. D. Stearns, *Adaptive Signal Processing*. Englewood Cliffs, NJ: Prentice-Hall, 1985.

[164] S. G. Wilson, *Digital Modulation and Coding*. Englewood Cliffs, NJ: Prentice-Hall, 1996.

[165] E. Wong, *Stochastic Processes in Information and Dynamical Systems*. New York: McGraw-Hill, 1971.

[166] P. M. Woodward, *Probability and Information Theory, with Applications to Radar*. Elmsford, NY: Pergamon Press, second ed., 1964.

[167] J. M. Wozencraft and I. M. Jacobs, *Principles of Communication Engineering*. New York: Wiley, 1965.

[168] W. W. Wu, *Elements of Digital Satellite Communication*. New York: Computer Science Press, vol. I, 1984.

[169] C. R. Wylie and L. C. Barrett, *Advanced Engineering Mathematics*. New York: McGraw-Hill, fifth ed., 1982.

[170] R. D. Yates and D. J. Goodman, *Probability and Stochastic Processes: A Friendly Introduction for Electrical and Computer Engineers*. New York: Wiley, 1999.

[171] J. H. Yuen (ed.), *Deep Space Telecommunications Systems Engineering*. New York: Plenum, 1983.

[172] R. F. Ziemer and R. I. Peterson, *Digital Communications and Spread Spectrum Systems*. New York: Macmillan, 1985.

[173] R. E. Ziemer and W. H. Tranter, *Principles of Communications*. Boston, Mass: Houghton Miflin, third ed., 1990.

Index

μ-law, 112
3 dB bandwidth, 215

a priori probabilities, 126
A-law, 112
acquisition, 301
adaptive DPCM, 122
adaptive prediction, 121
alphabets, 107
alternative mark inversion, 113
AM–PM conversion, 72
analog-to-digital converter, 111
aperture effect, 104
autocorrelation function, 24
automatic gain control, 229
average BER, 269
average probability of (symbol) error, 145

balanced frequency discriminator, 69
balanced modulator, 91
bandwidth-limited, 13
bit, 113
bit error rate, 135
block synchronization, 229
bluetooth, 304
Boltzmann, 36

capture effect, 83
carrier phase recovery, 229
CDMA and multi-user communications, 17
center of (probability) mass, 182
channel, 1
chip, 296
circuit switching, 8

clipping, 288
code word, 251
coding gain, 261
coherence bandwidth, 268
coherence time, 268
coherent, 194, 299
coherent communication, 185
combining, 272
communications, 1
compander, 112
compressor, 112
constant-envelope, 194
constraint length, 255
continuous-phase, 203
continuous-time random process, 22
correlation receiver, 175
correlative-level coding, 158
coset, 253
coset leader, 253
cross-correlation functions, 26
cyclic prefix, 280
cyclic redundancy check, 249

data aided, 229
data compression, 116
data networks, 17
decision feedback equalizer, 156
decision levels, 108
decision rule, 172
decision thresholds, 108
delay power spectrum, 268
delta modulation, 116
detection and estimation, 17
deviation ratio, 208
difference signal, 69
differential encoding, 114

differential PCM, 122
differential PSK, 194
digital communications, 17
digital hierarchy, 115
direct-conversion, 73
directional gain, 238
directivity, 238
discrete-time Fourier transform, 99
discrete-time random process, 22
dispersion, 135
Doppler power spectrum, 268
duo-binary encoder, 159
duo-binary signaling, 158

early-late gate synchronizer, 236
effective aperture, 238
effective isotropic radiation power,
    238
equal-gain, 272
equally probable, 142
equivalent noise temperature, 239
error-correcting/control codes, 17
error propagation, 159
error rate, 113
expander, 112
eye pattern, 157

fade, 266
fading, 266
fading channels, 265
false alarm, 142
fast fading, 268
figure of merit, 74
finite duration impulse response, 118
flat fading, 268
frame synchronization, 229
frames, 113
frequency selective, 268

Gaussian filtered MSK, 215
go-back-N, 250
Gold codes, 305
Gram–Schmidt (orthogonalization)
    procedure, 165

Hamming distance, 251

ideal Nyquist channel, 150
IEEE transactions on communica-
    tions, 291
information, 1
information theory, 17
inter-channel interference, 280
inter-symbol interference, 135, 280
interference-limited, 12
intermediate frequency, 72
inverse filter, 131

jointly strictly stationary, 23

$k \times n$ generator matrix, 252

likelihood function, 171
line coding, 158
linear, 251
linear FM signals, 93
linear prediction, 119
link budget, 236
link calculation, 236
link margin, 237
log-likelihood function, 172
low-density parity check codes, 16

M-ary, 194
m-sequence, 297
matched filter, 139
maximum a posteriori probability
    (MAP), 173, 230
maximum length code, 297
maximum likelihood sequence
    estimation, 209
maximum ratio combining, 274
mean-square error, 120
memoryless, 170
missing, 142
mobile communications, 17
modified Bessel function, 41
modulation, 33
multipath intensity profile, 268
multiple-input-multiple-output
    (MIMO), 271
mutually disjoint, 20

near-far, 306
noise enhancement, 131, 151
noise-quieting effect, 82
non-coherent, 194, 300
non-coherent matched filter, 221
non-coherent orthogonal modulation, 221
non-DA, 229
non-frequency selective, 268
non-stationary, 23
nonuniform quantizer, 108
NOT, 260
Nyquist criterion, 150
Nyquist rate, 101, 150

observation, 170
observation space, 173
OFDM, 17
offset QPSK, 198
optical communications, 17
optimum decision rule, 173
outage, 266

packet switching, 8
packets, 113
partial response, 158
passive, 5
peak-to-average-power ratio, 287
periodogram, 33
phase noise, 287
postcursors, 156
post-detection filter, 79
power control, 306
power gain of an antenna, 238
power-limited, 12
power spectra, 194
power spectral density, 31
precoding, 159
prediction error, 120
principles of mobile
    communications, 269
probability space, 19
processing gain, 296, 299
pulse-amplitude modulation, 102
pulse code modulation, 106
pulse-duration modulation, 105
pulse position modulation, 105

quadratic receiver, 221
quadrature-amplitude modulation
    (QAM), 194
quantization, 106
quantization error, 108
quantization noise, 108

radiation efficiency factor, 238
random process, 22
random variable, 20
Rayleigh fading channel, 267
receive diversity, 271
receiver, 1
regeneration, 116
Ricean fading channel, 267
roll-off factor, 157
rotationally invariant, 188

sample and hold, 103
samples, 99
sampling, 106
sampling period, 99
sampling theorem, 101
selective ARQ, 250
selective combining, 272
set partitioning principle, 16, 262
shot noise, 35
sigma–delta modulation, 118
signal constellation, 172
signal space, 162
signal-to-noise ratio, 73
spectrum analyzer, 93
spread, 268
square-law modulator, 90
state-transition, 255
stationary, 23
stationary process, 24
stereo multiplexing, 69
stop-and-wait ARQ, 249
strictly stationary, 23
sufficient statistics, 171, 233
sum signal, 69
superheterodyne receiver, 72
switching modulator, 90
symbol, 135
synchronization, 114, 148

synchronization and receiver
   design, 17
systematic code, 253

thermal noise, 36
threshold effect, 78
time average, 303
tracking, 301
training sequence, 156
translation invariant, 188
transmit diversity, 271
trellis diagram, 255
turbo codes, 16

uniform quantizer, 108
uniformly most powerful, 221
union bound, 182

Viterbi algorithm, 209, 257

white noise, 36
wideband communication, 292
Wiener–Hopf, 120
world wide web, 9

ZF equalizer, 151